D1188477

Futures Research and Environmental Sustainability

Theory and Method

Futures Research and Environmental Sustainability

Theory and Method

James K. Lein

CRC Press is an imprint of the
Taylor & Francis Group, an **informa** business

CRC Press
Taylor & Francis Group
6000 Broken Sound Parkway NW, Suite 300
Boca Raton, FL 33487-2742

Printed by CPI on sustainably sourced paper
Version Date: 20161019

International Standard Book Number-13: 978-1-4987-1660-4 (Hardback)

Library of Congress Cataloging-in-Publication Data

Names: Lein, James K., author.
Title: Futures research and environmental sustainability : theory and method / James K. Lein.
Description: Boca Raton, FL : Taylor & Francis Group, 2016.
Identifiers: LCCN 2016025913| ISBN 9781498716604 (hardback : alk. paper) | ISBN 9781498716628 (ebook)
Subjects: LCSH: Sustainability. | Environmental protection. | Forecasting--Research.
Classification: LCC GE196 .L45 2016 | DDC 338.9/27--dc23
LC record available at https://lccn.loc.gov/2016025913

Visit the Taylor & Francis Web site at
http://www.taylorandfrancis.com

and the CRC Press Web site at
http://www.crcpress.com

To Christine … here's to the future!

Contents

List of Figures

List of Tables

Preface

Human impact on the environmental system has been a topic of interest to me since my undergraduate studies in geography. Much of my professional career, whether in the classroom or in research, has been dedicated to this theme. This book focuses more on real-life examples from my experience to explore this topic and share my discoveries. The subject of human–environment interaction, as we know, is complex and complicated. The published studies documenting the wide range of effects human activities have had on the environmental system display enormous range and variety with substantial contributions from nearly all academic disciplines. The evidence gleaned from these studies details the deleterious consequences of past actions and forms an extensive list of the "what not to do" that, on occasion, informs the choices and decisions we make. Although looking into the past can be informative, the connection to the future, where human impact is realized, has not been actively pursued. Aside from the nod given to the "future generation" ethic that is carefully placed in the front matter of environmental policy directives, the explicit study of the future from an environmental perspective is not actively engaged.

Several years ago I stumbled upon the field of futures research/future studies. In that literature I saw potential for this subject matter to provide important insights and methods to better understand human impact on the environment in a more relevant way. If we are committed to the notion of environmental stewardship with an eye toward those future generations that will inherit the impacted environmental system, there is a need to identify tractable methods whereby human progressions can be examined outward into the future and evaluated in relation to a dynamic environmental setting. Merging the two subject areas—human–environment interaction and futures research—seems like a logical place to begin this search. Synthesis, however, requires a focal point around which theory and method can be blended. Sustainability serves as that point of convergence where human activities can be examined in relation to their capacity to be supported over the long term. Planting a tree in the name of sustainability may be a wonderful thing provided the tree lives. The uncertainties of the future and the desire to know what may seem to be unknowable informs our tree planting experience, and so I set out to bring together the concepts and methods of futures research with the developing body of sustainability theory to explore human–environment interaction in its proper context. The aim of this book is to present an accessible digest of background and methods to guide the assessment of current actions and their long-term environmental sustainability.

Author

James K. Lein is a professor of geography at Ohio University in Athens, Ohio. He specializes in the environmental sciences and the application of remote sensing and geographic information systems in environmental planning, impact assessment, and regional landscape analysis. A product of the San Francisco Bay Area, he earned a BA in geography at San Francisco State University, an MA in geography at San Jose State University, and a PhD in geography at Kent State University. Professor Lein has served on the faculty at Ohio University since 1989, where his teaching focused on environmental geography, planning, physical geography, and the geographic techniques.

Professor Lein has spent his career in applied geographical research. His research interests currently focus on the monitoring of land use/land cover change, land resource science, and the integration of remote sensing and geographic information systems in environmental assessment and decision-making. His major publications include *Environmental Decision Making: An Information Technology Approach* and *Integrated Environmental Planning and Environmental Sensing: Analytical Techniques for Earth Observation.*

Professor Lein has also published in journals such as *Sustainability, Papers in Applied Geography, Environmental Practice, GeoScience and Remote Sensing, Environmental Technology and Management, International Journal of Remote Sensing, GeoCarto International, International Journal of Environmental Studies, Applied Geography, Journal of Environmental Management,* and in numerous conference proceedings.

Introduction

There have been several themes that have run constant in the practice of environmental management and planning for the past four decades. Sustainability takes its place among these and has quickly become a focal concern that is actively reshaping environmental policy and decision-making. So pervasive is this concept that the notion of sustainability has been stretched into the far corners of society that it risks losing meaning. To why I eat a particular breakfast cereal to the manner by which decisions are made to combat global climate change, the word sustainability is injected somewhere to capture our attention or direct our behavior. Sustainability is one of those wonderful ideations that generate positive agreement because it invokes commonsense beliefs. However, sustainability also produces considerable frustration when attempts are made to move this idea from the conceptual to the more pragmatic. Defining, describing, quantifying, knowing sustainability, like many other concepts that "bridge" human and the environment, remain a challenge. Without meeting this challenge and crafting the good idea of sustainability into a more actionable framework, it runs the risk of becoming irrelevant.

Relevance is important only because it implies that there is a body of evidence to support the use of sustainability as an actionable metric. Relevance is important also because it suggests that this metric is sufficiently meaningful to guide decision-making. However, to this point in the search for sustainability, there is no "yard stick" to measure and compare actions and outcomes. There is a developing body of work that is attempting to define sustainability according to more precise criteria and produce methods of investigation that are hoped will facilitate its representation. Representation, as if sustainability is something that can be pointed to, introduces a temporal and spatial dimension that has not been well examined. When is something sustainable and where is something sustainable are two critical questions if sustainability is to become something more than a label. Plans and policies designed to yield something that is "sustainable" need to address the where and when. Neither question is easy to answer and poses an additional challenge to those seeking a more scientific approach that promises to move the concept of sustainability forward.

The where and the when are two critical features of the sustainability question that serve as the motive for this book. While there is no shortage of books and monographs that explore the topic of sustainability, the where and the when, as I discovered, are perplexing questions that have not been widely explored. The complexity of what the ideal of sustainability entails and the difficulties that surround its absolute and consistent meaning introduce all manner of pitfalls that explain why this gap in the literature exists. Nevertheless, it is a gap that we need to find ways to fill in order to complete that bridge between human activities and decisions that find themselves interacting in unanticipated and problematic ways with environmental processes. With the when and the where as motivation, the goal of this book is to understand and explore the challenges of (1) presenting sustainability as a more actionable or practical concept and (2) identifying approaches that might offer useful assistance in addressing the temporal and spatial representation of sustainability. The underlying premise of this book is that sustainability is a state realized in the future. In that future there is a geographic arrangement of society and economy that agrees with its environmental setting. This future perspective introduces a little-examined subject area that can lend significant content to the sustainability challenge: Futures Research. With so much of the relationship between human actions and environmental processes contingent on the futures, methods that might offer insight to "know" tomorrow and explore the pathways human decisions may take as they unfold over time is an essential, and missing, element in the search for sustainable outcomes.

Integrating a futures research perspective with the existing body of knowledge regarding the topic of sustainability, sustainable development, and the environment begins in Chapter 1. Here, the art and science of sustainability is examined from its early roots to the issues that influence its many definitions. This chapter introduces the conceptual models that characterize the dimensions of sustainability and draws specific attention to the environment and the foundation for the study and assessment of this concept. Chapter 2 builds on this discussion and introduces the topic on assessment. Here the procedural aspects that develop into a method are examined based on the requirements that can become part of the environmental assessment process. Focus in this chapter is given to the identification of indicators and metrics that can be incorporated into an assessment framework and employed to monitor progress toward specific sustainability targets.

Assessment and the inherently "future"-oriented nature of sustainability decision-making operate in an environment punctuated by uncertainty. Chapter 3 examines the role and importance of uncertainty through an examination of its definitions and the manner by which it influences prediction. This treatment of uncertainty with an eye toward prediction provides a transition to Chapter 4 where the foundation principles of futures research are presented. Tracing the origins of futures research in brief takes the discussion into a treatment of forecasting, prospective analysis, and visioning to suggest ways to ascertain the "when is sustainable" question. The methods of futures research are reviewed in Chapter 5. The qualitative and quantitative approaches available to apply judgment and forecasting to understand possible futures are described in relation to the question of environmental sustainability. The geospatial examination of environmental sustainability is the subject of Chapter 6. Here the question of "where is sustainable" is addressed through a discussion of spatial models and their implementation using a geographic information system.

After these technical discussions, we return to the concept of a scenario first introduced in Chapter 2. Here, an expanded treatment of this essential ingredient of futures research and forecasting is provided. This includes consideration on how to craft scenarios to support forecasting, manage uncertainty, and use scenarios as a planning and visioning device. The final chapter of this book considers the question of change the topic of resilience as a planning focus. From this discussion the chapter moves to a treatment of the policies and procedural steps involved in implementing sustainability strategies and concludes with an examination of "best practices" currently used and advocated to guide development toward more environmentally sustainable outcomes.

As can be seen from this synopsis, I adopt an environment bias with regards to the concept of sustainability. My rationale for placing the environment in this central role is not to diminish the importance of the economic or social aspects that contribute to the tripart definition of sustainability, but rather to recognize the importance of the environmental system as the supporting foundation on which all other activities depend and interact. Thus, a degraded environment cannot sustain economic or social interactions in the long term—the place where sustainability is realized. Therefore, when exploring the degree to which human actions can be sustained, the environment takes precedence. Development plans and agendas that ignore the environment cannot successfully promote economic expansion or social well-being. The environmental focus also provides a more accessible set of conditions to frame assessment, decision-making, and forecasting. With these considerations in mind, this book was developed to engage these three topics and provide a synthesis of information from several fields of study that lend support to the understanding and evaluation of sustainability. The target audience reflects a wide mix of environmental professionals that are working to make sustainability a real and achievable target—from those in planning and resource management to others focused on the natural resource issues that are embedded in the idea of a sustainable environment. This book is also intended to serve as a first exposure to the methods and procedures for assessing and forecasting human activities in relation to the environment to undergraduate students in the emerging science of sustainability, whether in traditional academic programs such as environmental studies and geography or the newly minted disciplines that have sustainability in their titles.

Curran, 2009). Facing these challenges following traditional planning models has been criticized as inconsistent with emerging realities (Lélé, 1991). Revisiting existing methods of environmental decision-making has been advocated for decades and each new installment has attempted to incorporate ecological ideals and principles into the development, policy, and planning arena (Lein, 2003). Within the last decade, the concept of sustainability has reemerged as one more opportunity to think broadly regarding the form and consequence of human actions and their interaction within the larger environmental system. Presently, "sustainability" or "sustainable thinking" has taken its place among the list of decision-centric models to address environmental concerns. The precise meaning of the term, what specific it explains, and how this concept may be applied to direct human activity in relation to the environmental system remain elusive (Marshall and Toffel, 2005). In the section that follows, the definitions and theories that underpin the concept of sustainability are examined.

1.2 Sustainability Theory

The history of "environmentalism" does much to inform the evolution of the concept of sustainability. Consistent observations describing the aftermath of the destructive exploitation of the environment in the name of progress has been documented since the fifth century BC, introducing a pattern carried forward through the Industrial Revolution to the present (Du Pisani, 2006). However, what presently indentifies as sustainability thinking seems to have emerged out of the forests of Western Europe in the early seventeenth century (Du Pisani, 2006). The concern during this period was the practice of timber harvesting and the need to balance tree removal with the preservation of forest lands to ensure forest regeneration and ultimately the long-term viability of forest resources. According to Du Pisani (2006) and Wiersum (1995), the concept of sustainability was formulated in the German forestry literature as the principle of *Nachaltigkeitsprinzezip*. Several central themes developed from this principle are embedded in the current insanitation of sustainability as used today. Chief among these is the explicit focus on future generations, the wise use of resources, and the maintenance of productive environments (Wiersum, 1995). Based on these ideals, sustainability is, in essence, a hypothesis concerning the actions taken by society and the capacity of a neutral environment to maintain a level of resilience when those actions define a damaging or disruptive force. As a hypothesis, sustainability carries the implicit assumption that a level of human activity can be defined whereby society can continually "harvest" the environment and the environment will continue to function unperturbed.

At its root, sustainability theory encapsulates a set of interconnecting maxims that have influenced environmental management since the 1950s (Kidd, 1992):

- *Ecological carrying capacity*: Carrying capacity emerged from the biological sciences, where it explains the relationship between the resource base, the assimilative and restorative capacity of the environment, and the biotic potential of a species (Lein, 2003). When applied in an environmental or resource management context, carrying capacity represents an assumption that a level of human activity, expressed as a pressure, exists that a region can support at an acceptable quality of life without engendering significant environmental damage. Alternatively, carrying capacity seeks to define the maximum ability of the environment to continuously provide resources at a level required by a given population.
- *Limits to growth*: Limits to growth attempts to articulate the tension that develops between population growth pressures, economic expansion, and environmental degradation. The limits to growth argument maintain that economic and population growth pressures are direct causes of environmental decline. Therefore, continued population growth and exploitive forms of economic development cannot be sustained. Degraded environmental conditions will impose limits through a declining resource base on this further growth. Those limits will ultimately retard further expansion.
- *Global teleconnections*: Although the term "teleconnection" is best understood in the atmospheric sciences, when placed into an environmental context, it directs attention to the biosphere and the role human activities have in generating anomalous patterns that alter large-scale global processes. The biosphere, describing the global sum of all ecosystems, characterizes the zone of life on earth. The increasing scale and intensity of human activities serve as agents of change, extending their influence to the global scale and impacting societies well beyond the locales where these actions originate, often with cross-generational implications.
- *Appropriate technology*: The ideology expressed under the notion of appropriate technology encourages a reexamination of the manner by which scientific knowledge is applied and how the machinery and processes are developed to make these technologies impart an influence on society and the environment. Technology, in the broadest sense, represents the range of ways by which social groups provide for themselves and the material fabric that characterizes their way of life. Technology also serves to define the "end point" of the resources devoted to maintain a given social system as well as a crude measure of a society's level of economic development.

Identifying the level of technology that is balanced with the ecological system and its supportive capacity is the goal.

- *Soft-path development*: Growth strategies typically rely on the creation of new infrastructure and investments. Traditionally the growth model is described as capital intensive, centralized, resources intensive, resource depletive, and resource degrading. Soft-path strategies focus on capital-extensive, small-scale, and decentralized forms of development that reinforce conservation, life-style opportunities, and social equity.

Blending these foundational concepts into a cohesive statement that connects to the ideal of sustainability centers on the unifying theme of ecological dependency. The ecological model, defining a system of components organized into a self-regulating arrangement of living and nonliving elements, implies that sustainability is an action-centered supposition that frames the conflict between environmental processes and the status of social, economic, and ecological connections. Sustainability theory seeks to explain the reciprocal exchange between degraded environmental processes resulting from human activities and the vulnerability of those activities to a degraded environment. Therefore, the capacity to maintain an entity, outcome, or process over time and the inherent nature of those features to degrade the "capacity to maintain" underscores the concept of sustainability (Figure 1.2). The circular nature of human–environment interaction suggested by this theory further accentuates the need for continuity within a human system dependent upon unaltered environmental processes for its vitality and long-term survival. While the fundamental ideals that form the concept of sustainability seem fundamental, a clear formulation of this theory has been frustrated by the problem of definition (Cairns, 2004; Holden, 2010; Kuhlman and Farrington, 2010).

1.3 The Definition Problem

A theory is essentially a proposed explanation whose status is still conjectural, subject to examination and often standing in contrast to more well-established ideas or principles. Sustainability, as a theory of human organization, suffers from a degree of linguistic ambiguity that hampers both the formulation of method and clarity of focus (Childers et al., 2014). There are many ways to define sustainability. Perhaps the most widely used definition can be traced to the United Nation's World Commission on Environment and Development. Since the publication of this report, the concept of sustainability has broadened and become used interchangeably with terms such as sustainable development and environmental sustainability.

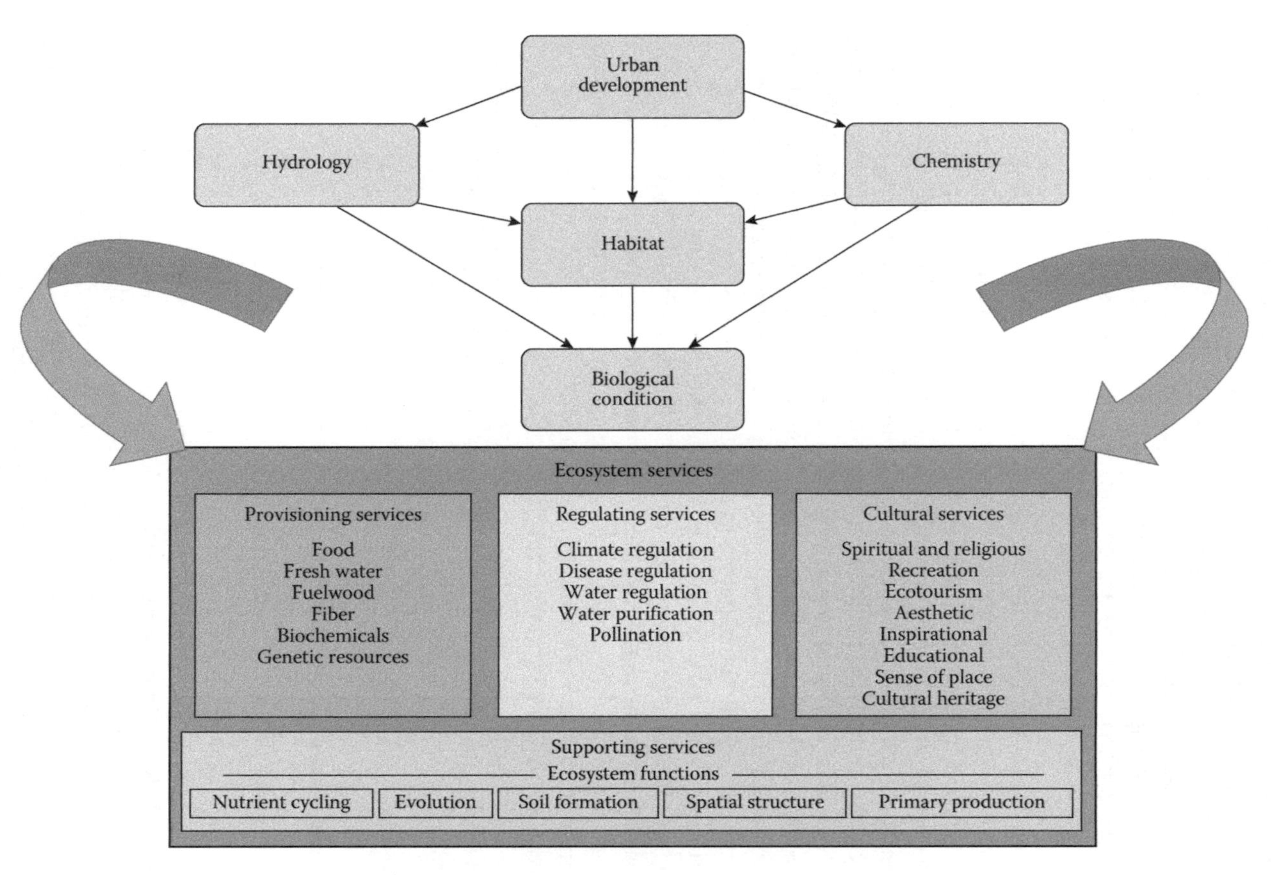

FIGURE 1.2
The coupled system of environmental process and ecosystem support services.

Indeed, it has been suggested that sustainability has become a contemporary "buzzword" too often relegated to the murky waters of policy discourse (Redclift, 2005; Tovey, 2009; Wass et al., 2011).

The widely cited World Commission definition characterizes sustainability as "...development that meets the needs of the present without compromising the ability of future generations to meet their own needs...." This definition identifies the fundamental concerns that must be reconciled in order to realize sustainability: the tension between development and environment but has been criticized as overly broad (Kroop and Lein, 2012). More recent attempts to produce a succinct definition of sustainability have been cataloged by Murcott (1997) and Kidd (1992). Definitions beginning from the mid-1970s, predating the World Commission's offering, include

- Living within the self-perpetuating limits of the environment
- Development that is likely to achieve lasting satisfaction of human needs and improvement of the quality of life
- A pattern of social and structural economic transformations that optimizes the economic and social benefits available in the present without jeopardizing the likely potential for similar benefits in the future
- An ability to maintain some activity in the face of stress
- The development and management of natural resources to ensure or enhance the long-term productive capacity of the resource base and improve the long-term wealth and well-being derived from alternative resource use systems, with acceptable environmental impacts
- A society based on a long-term vision in that it must foresee the consequences of its diverse activities to ensure that they do not break cycles of renewal
- Meeting human needs without compromising the health of ecosystems
- Transforming our ways of living to maximize the chances that environmental social conditions will indefinitely support to enhance human security, well-being, and health

The range of definitions, while instructive, runs the risk of confusing meaning and has contributed to the overuse and misuse of the concept (Károly, 2013). To improve meaning and restrict interpretation to a more actionable concept, a simple four-part typology has been advanced (Sutton, 2004a):

1. Definitions based on essence where sustainability explains the ability to "sustain" something, such that sustainable development is development that can be maintained and sustaining development is development that sustains something

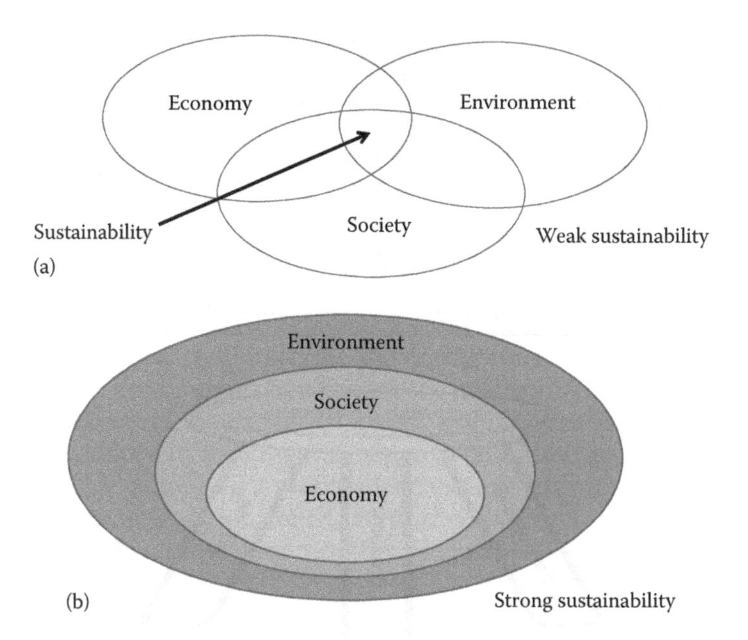

FIGURE 1.4
Integrative models of sustainability identifying (a) overlapping elements and (b) nested elements of the concept.

Regardless of conceptual design, the models demonstrate one unifying theme: that sustainability must be addressed as a complex system of linkages and interdependencies. Therefore, while further elaboration of the conceptual aspects of sustainability is likely, it is possible at present to identify four fundamental considerations that are critical to the formulation of any sustainability plan or policy agenda (Waas et al., 2011):

1. *The normality principle*—recognizing that sustainability is socially constructed, subjective, and influenced by normative choices
2. *The equity principle*—pertains to the ideal of fairness and justice articulated in relation to intergenerational, intragenerational, geography, procedural, and interspecies obligations and responsibilities
3. *The integration principle*—realizing that sustainability should harmoniously blend into various social and institutional development objectives as well as the environment
4. *The dynamism principle*—recognizing that sustainability is a process of directed change and should be regarded as a continuous search for an elusive equilibrium in a dynamic setting

Models and definitions help to focus and clarify planning and policy efforts aimed at creating strategies to achieve sustainability, yet models alone merely illustrate the paradigm. How sustainability is approached in an actionable program depends on the degree to which an object incorporates sustainability ideals. In the next section, two prominent modes of sustainability are examined.

1.5 Weak versus Strong Sustainability

In practical terms, the sustainability paradigm separates into two contrasting styles or perspectives: weak and strong. The definition of either term hinges on an understanding of the more specific interactions between the environment and the economic system. Weak sustainability defines the nonhuman world in relation to its economic value. This economic expression of sustainability is based on the premise that as long as diminishing natural capital stocks that explain the resource base are being replaced by gains in the human-made stock, the totality of capital available as support will remain constant (Liu, 2009). This consistency will allow the current level of consumption to continue in a sustainable pattern. This balance sheet view adopts a human-centered view of the relationship between people and the environmental system that

- Emphasizes a growth-oriented approach to economic development
- Undervalues the need for changing societal demands on the resource base
- Perpetuates the mindset that nature is a collection of natural resources that can be controlled by humans

By contrast, strong sustainability is incorporated on a more considered view of the human–environment interaction. Advocates of strong sustainability recognize the need for small-scale, decentralized forms of development in order to create social and economic systems that are more ecologically balanced and less destructive. Strong sustainability does not make allowances for the substitution of human capital for natural capital as is the case with weak sustainability. Rather, according to this paradigm, products created in the human system cannot replace lost or compromised natural capital found in ecosystems. In some respects, strong sustainability can be conceptualized as a series of thresholds that cannot be exceeded. Human development and the potential impact it may generate must, therefore, fall within thresholds established through the interplay between social and political preference with careful consideration given to the resilience and recovery strategies of a natural system under stress. The objective of strong sustainability, therefore, is to protect the environmental system, not only because of its significance to future human progress, but because the natural system enjoys biotic rights above those of society (Williams and Millington, 2004). In this context, strong sustainability capitalizes on the principles for environmental stewardship to guide the sustention of ecosystem services.

In the decision-making arena, both weak and strong sustainability play important roles in establishing policy objectives that include

- Limiting the scale and intensity of human activities within the environmental system to a level that does not exceed the "carrying capacity" of natural capital
- Focusing technological development on increasing the efficiency of resource use
- Managing the sources and sink functions of renewable natural capital to not exceed rates of regeneration or the assimilative capacity of the environment
- Balancing the rate of non-renewable resource exploitation with the rate of creation for renewable substitutes

Rather than polar opposites, weak and strong sustainability explains a continuum that describes how society views and uses the environment. Whether a weak, strong, or an intermediate direction is adopted, the choice between these conditions develops from a mixture of scientific, ethical, pragmatic, and political deliberation. From this vantage point, the weak–strong continuum aptly demonstrates that there is no purely object- or value-free way to understand or implement sustainability.

1.6 Embracing the Environmental

Although there may be numerous ways to couch a definition of the term sustainability, the sustainability paradigm displays an undeniable environmental focus that is derived primarily from an ecological or environmental perspective (Curran, 2009). The tendency to view sustainability through an environmental lens is not to diminish social, economic, urban, or other considerations, rather it recognizes that strategies developed to realize a sustainable solution all hinge on the overarching theme of environmental performance. By maintaining ecosystem integrity and health, economic, social, and urban systems can be sustained. As the environment stands as a substantial constraint to human progress in the modern era, the maintenance of life-support systems becomes a prerequisite for functional social, economic, and urban systems (Goodland, 1995). Placing the

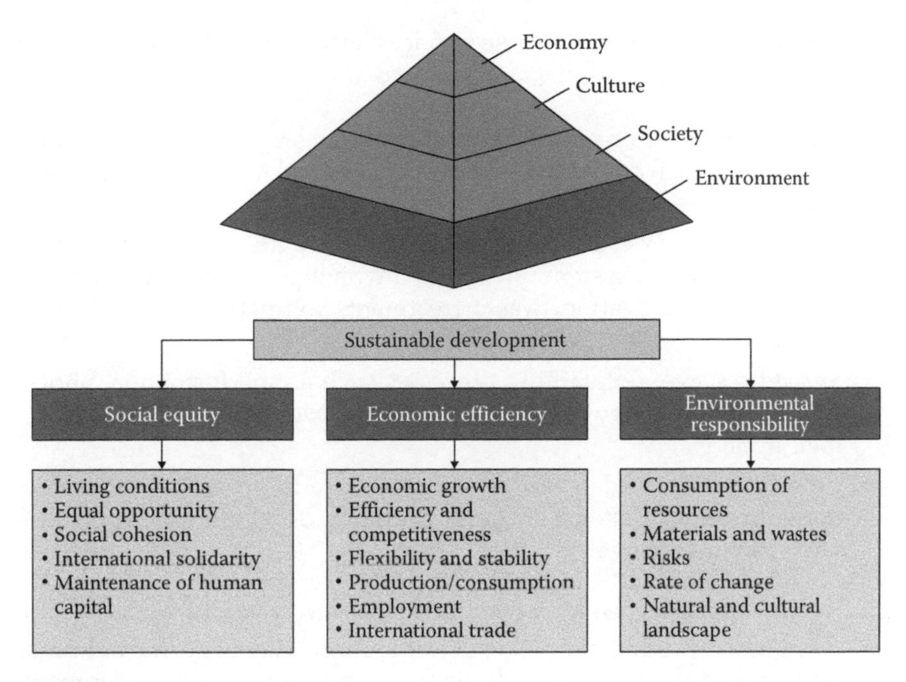

FIGURE 1.5
The environmental base supporting the sustainability pyramid.

environment as the foundation of the sustainability pyramid suggests a more realistic model and gives emphasis to the importance of a productive environment as a necessary condition (Figure 1.5). This model further illustrates the inherent connectedness between all elements of the system; implying that a sustainable economy depends on a sustainable flow of energy and materials such that a sustainable society is embedded in that economy and the urban world, occupying the top tier in this model, requires long-term stability of all three elements below it to ensure its sustainability.

Environmental sustainability forms out of the set of constraints on the activities that govern the scale of resource appropriation and the process of waste assimilation. The importance of this connected tier structure, therefore, is the underlying theory that environmental sustainability is essentially a natural science construct and obeys biophysical principles to define appropriate conditions of balance, resilience, and interconnectedness between human society and the supporting ecosystem (Goodland, 1995; Morelli, 2011). While a sustainable environment need not depend on the existence of either society or economy, the opposite is not the case. Therefore, a crucial characteristic of a sustainable environmental system is its dependency on the flow of ecosystem services that cycle upward through the pyramid model. Those services have been outlined by Morelli (2011) and include

- *Provisioning services*—identifying the products society obtains from ecosystems such as food, fiber, genetic resources, biochemical, freshwater, and energy
- *Regulating services*—defining those benefits gathered through the regulation of ecosystem processes involving air quality regulation, water purification, waste treatment, pollination, and hazard regulation
- *Support services*—explaining processes such as soil formation, photosynthesis, primary production, nutrient cycling, and the hydrologic cycle
- *Cultural services*—describing the nonmaterial benefits humans derive from ecosystems through cognitive development, amenity, and spiritual renewal

Recognizing that ecosystems range in complexity from relatively undisturbed to landscapes increasingly subject to more intense forms of human modification and management, directing attention to the "environment" when exploring sustainability outcomes, concentrates on those land types where human use and impact is "significant" and where human dependency is greatest. From this approach, environmental sustainability becomes the purposeful task of minimizing the disturbance and degradation of ecosystems while maintaining their productivity. In practical terms, this directed form of ecological preservation seeks to optimize land, energy, and material use within the context of human development and occupation of the landscape. Although the concept of development continues to evolve, with environmental sustainability as the goal, sustainable development demands the adoption of an economic model that does not foster the expansion of matter and energy "throughputs" beyond the regenerative and absorptive capacities of the local environmental system (Goodland and Daly, 1996; Newman, 2005).

Development, from a sustainability perspective, contrasts markedly with the notion of growth. Growth typically implies an increase in size. Drawing from a term referred to in the environmental planning literature as the "Brontosaurus Principle," increase in size may be manageable up to a point, whereafter, continued growth becomes a liability (Baldwin, 1985; Miller, 1998; Lein, 2003). The Brontosaurus analogy affirms the presence of thresholds in the environment. When growth exceeds those thresholds, it becomes unsustainable. Sustainable development is therefore a dynamic and continual process whereby communities anticipate and accommodate the needs of current and future generations in ways that are assumed to be reproducible and balance the local social, ecologic, and ecological systems

FIGURE 1.6
Sustainability balanced by societal motivations and environmental risk.

(Berke, 2002). The presence of thresholds (real or theoretical) informs this process and establishes a framework to approach the challenge of sustainability assessment (Figure 1.6).

1.7 Summary

The sustainability construct is not a new idea. The principles of balance between human actions and the environmental system, the capacity of future generations to enjoy the resources and support services that presently support current development patterns and to respect the constraints imposed by the natural systems that act to moderate future human activities are embedded a list of maxims that encapsulate sustainability thinking. Regardless of how sustainability may be defined, addressing this need for balance with a deeper commitment to the very long term requires addressing a complex system of linkages and interdependencies. Continual growth and economic development rest in the ability to achieve a requisite fit with the ecosystem services on which that growth depends. By embracing the environment as the central, and critical, "pillar" of the sustainability question, recognizing the dynamic nature of the environmental setting in which development occurs, and employing a considered understanding of the thresholds that exits with that setting, a more informed process of sustainability planning and assessment can be realized.

References

Baldwin, J. H. (1985). *Environmental Planning and Management.* Westview Press, Boulder, CO.

Berke, P. R. (2002). Does sustainable development offer a new direction for planning? Challenges for the twenty-first century. *Journal of Planning Literature, 17*(1), 21–36.

Bond, A. J. and Morrison-Saunders, A. (2009). Sustainability appraisal: Jack of all trades, master of none? *Impact Assessment and Project Appraisal*, *27*(4), 321–329.

Cairns Jr., J. (2004). Will the real sustainability concept please stand up. *Ethics in Science and Environmental Politics*, *22*, 49–52.

Childers, D. L., Pickett, S. T., Grove, J. M., Ogden, L., and Whitmer, A. (2014). Advancing urban sustainability theory and action: Challenges and opportunities. *Landscape and Urban Planning*, 125, 320–328.

Curran, M. A. (2009). Wrapping our brains around sustainability. *Sustainability*, *1*(1), 5–13.

Du Pisani, J. A. (2006). Sustainable development–historical roots of the concept. *Environmental Sciences*, *3*(2), 83–96.

Goodland, R. (1995). The concept of environmental sustainability. *Annual Review of Ecology and Systematics*, *26*, 1–24.

Goodland, R. and Daly, H. (1996). Environmental sustainability: Universal and non-negotiable. *Ecological Applications*, *6*(4), 1002–1017.

Goudie, A. S. (2013). *The Human Impact on the Natural Environment: Past, Present, and Future*. John Wiley & Sons, New York.

Gowdy, J. M. (1994). Progress and environmental sustainability. *Environmental Ethics*, *16*(1), 41–55.

Guha, R. (2014). *Environmentalism: A Global History*. Penguin, London, U.K.

Holden, M. (2010). The rhetoric of sustainability: Perversity, futility, jeopardy? *Sustainability*, *2*(2), 645–659.

James, P. (2014). *Urban Sustainability in Theory and Practice: Circles of Sustainability*. Routledge, New York.

Jay, S., Jones, C., Slinn, P., and Wood, C. (2007). Environmental impact assessment: Retrospect and prospect. *Environmental Impact Assessment Review*, *27*(4), 287–300.

Károly, K. (2013). Rise and fall of the concept sustainability. *Journal of Environmental Sustainability*, *1*(1). http://doi.org/10.14448/jes.01.0001.

Kenny, M. and Meadowcroft, J. (2002). *Planning Sustainability*. Routledge, New York.

Kidd, C. V. (1992). The evolution of sustainability. *Journal of Agricultural and Environmental Ethics*, *5*(1), 1–26.

Kuhlman, T. and Farrington, J. (2010). What is sustainability? *Sustainability*, *2*(11), 3436–3448.

Lein, J. K. (2003). *Integrated Environmental Planning: A Landscape Synthesis* (228p). John Wiley & Sons, Oxford, U.K.

Lélé, S. M. (1991). Sustainable development: A critical review. *World Development*, *19*(6), 607–621.

Liu, L. (2009). Sustainability: Living within one's own ecological means. *Sustainability*, *1*(4), 1412–1430.

Lozano, R. (2008). Envisioning sustainability three-dimensionally. *Journal of Cleaner Production*, *16*(17), 1838–1846.

Ludwig, D. (1993). Environmental sustainability: Magic, science, and religion in natural resource management. *Ecological Applications*, *3*(4), 555–558.

MacKenzie, D. (2012). The limits of growth revisited. *New Scientist*, *213*(2846), 38–41.

Mannion, D. A. M. (2014). *Global Environmental Change: A Natural and Cultural Environmental History* (375p). Routledge, New York.

Marshall, J. D. and Toffel, M. W. (2005). Framing the elusive concept of sustainability: A sustainability hierarchy. *Environmental Science and Technology, 39*(3), 673–682.

Martens, P. (2006). Sustainability: Science or fiction? *Sustainability: Science, Practice, and Policy, 2*(1).

McHarg, I. L., and Mumford, L. (1969). *Design with Nature* (pp. 7–17). American Museum of Natural History, New York.

Mebratu, D. (1998). Sustainability and sustainable development: Historical and conceptual review. *Environmental Impact Assessment Review, 18*(6), 493–520.

Miller, G. (1998). *Living in the Environment*. Wadsworth Publishing, Belmont, CA.

Mitchell, B. (2013). *Resource and Environmental Management*. Routledge, New York.

Morelli, J. (2011). Environmental sustainability: A definition for environmental professionals. *Journal of Environmental Sustainability, 1*(1), 19–27.

Morelli, J. (2013). Environmental sustainability: A definition for environmental professionals. *Journal of Environmental Sustainability, 1*(1). http://doi.org/10.14448/jes.01.0002.

Murcott, S. (1997). Definitions of sustainable development. In *Proceedings: AAAS Annual Conference, IIASA Sustainability Indicators Symposium*, Seattle, WA (pp. 87–93).

Newman, P. (2005). Sustainability assessment and cities. *International Review of Environmental Strategies, 5*(2), 383–398.

Redclift, M. (2005). Sustainable development (1987–2005): An oxymoron comes of age. *Sustainable Development, 13*(4), 212–227.

Redman, C. L. (1999). *Human Impact on Ancient Environments*. University of Arizona Press, Tucson, AZ.

Rosner, W. J. (1995). Mental models for sustainability. *Journal of Cleaner Production, 3*(1–2), 107–121.

Ruddiman, W. F. (2013). The Anthropocene. *Annual Review of Earth and Planetary Sciences, 41*, 45–68.

Simmons, I. G. (1993). *Environmental History: A Concise Introduction*. Blackwell, Oxford, U.K.

Spangenberg, J. H. (2004). Reconciling sustainability and growth: Criteria, indicators, policies. *Sustainable Development, 12*(2), 74–86.

Sutton, P. (2004a). What is sustainability. *Eingana, the Journal of the Victorian Association for Environmental Education, 17*, 1–7.

Sutton, P. (2004b). A perspective on environmental sustainability. *Paper on the Victorian Commissioner for Environmental Sustainability*, Melbourne, Australia. http://www.green-innovations.asn.au/A-Perspective-on-Environmental-Sustainability.pdf (accessed Oct 1, 2016).

Thomas, W. L. (1956). *Man's Role in Changing the Face of the Earth* (pp. 10–13). University of Chicago Press, Chicago, IL.

Todorov, V. I. and Marinova, D. (2009a, July). Models of sustainability. In: *18th World IMACS Congress* (pp. 1216–1222), Cairns, North Queensland, Australia.

Tovey, H. (2009). Sustainability: A platform for debate. *Sustainability, 1*(1), 14–18.

Turner, B. L. (Ed.). (1990). *The Earth as Transformed by Human Action: Global and Regional Changes in the Biosphere over the past 300 years*. Cambridge University Press, Cambridge, U.K.

Waas, T., Hugé, J., Verbruggen, A., and Wright, T. (2011). Sustainable Development: A Bird's Eye View. *Sustainability, 3*(10), 1637–1661.

White, M. A. (2013). Sustainability: I know it when I see it. *Ecological Economics, 86,* 213–217.

Wiersum, K. F. (1995). 200 years of sustainability in forestry: Lessons from history. *Environmental Management, 19*(3), 321–329.

Williams, C. C. and Millington, A. C. (2004). The diverse and contested meanings of sustainable development. *The Geographical Journal, 170*(2), 99–104.

2

Assessing Environmental Sustainability: The Challenge

Environmental assessment has been an integral component of resource management and related panning activities for well over five decades (Jay et al., 2007; Glasson et al., 2013). The objective of any assessment program aimed at the land management problem is twofold: (1) incorporate unique environment considerations into a decision-making process and (2) monitor progress toward specific land management or developmental goals. When performing an assessment, establishing the objectives, deriving the appropriate criteria, acting on the findings, and informing the decision-making process are essential ingredients to developing a complete understanding of the decision to be made and its possible outcome(s). With assessment focused on the environment, it becomes necessary to also develop a complete understanding of the physical and human conditions that define the landscape. That understanding highlights the suite of factors that influence the decision to be made, its probable and possible effects, and the capacity of the environmental system to support those effects. Environmental sustainability as an outcome of a decision process presents significant challenges to the assessment process. In this chapter, those challenges are explored and the approaches available to guide the assessment of environmental sustainability examined.

2.1 The Assessment Question

Assessing progress toward the goal of environmental sustainability implies an ability to ascertain the environmental consequences of specific policies or proposals in relation to the environmental systems and its capacity to support changing conditions. Consequence in the context of sustainability relates primarily to the environmental system's ability to support the socio-economic and biophysical processes active within a defined regional setting. Assessment examines the degree to which human activities can remain supported in perpetuity. Several ideas converge on this question, from those related to traditional theories of environmental impact to the broader ideals of optimality as a social system progressed toward a development model that is aligned with the concepts of sustainability. Impact on the environment and optimality of choice, together with a collection of evaluative criteria, provide a framework to guide judgment and decision-making. As with any decision-making process, there is a desire to evaluate the ramifications of choice prior to making the choice selection and sustainability is no exception. Examining the scope and consequence of an action as it unfolds underscores the importance of proactive thinking in the process and emphasizes the complexity of choice when attempting to consider precisely what actions are "sustainable" given the future-oriented connotation of the term.

Organizing thought and forming a strategy to craft an assessment program takes the sustainability question and separates it into three main components: (1) an environmental, which concerns the nature of impacts to or effect on ecological process, (2) a strategy that examines the direction and intent of a policy or plan articulating development agendas, and (3) a sustainability link that considers the long-term fit or balance to be achieved in relation to the desired outcome (Figure 2.1). This three-tiered approach to assessment recognizes that the complexity surrounding the question of environmental sustainability may be reduced to a set of conditions that, when viewed collectively, give insight into the key decision points that direct choice and influence change in the landscape system. Identifying key decision points and illuminating the pathway we may take when decisions are made, rests at the core of assessment. Conceptually, a decision may branch and dissolve into a series of "what if" questions as the attempt is made to address the larger uncertainty of "is it sustainable?" The art and science of assessment by identifying and quantifying the benefits as well as the problems attributable to a course of action presents the means to address this vexing question.

Assessment is a widely used term, but often poorly understood. Various explanations have been offered to help define its role with a desire to

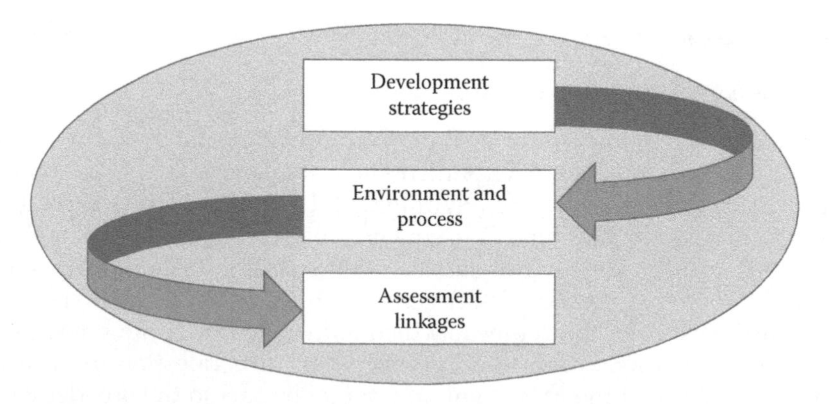

FIGURE 2.1
Linking strategies with processes to frame assessment activities.

produce environmentally sustainable outcomes (Bond and Morrison-Sanders, 2011; Bond et al., 2012; Gibson et al., 2013). Peering into the unknown is essential to this end, not only to guide the development of individual projects or directives, but also to monitor and continuously improve existing programs. Assessment also provides evidence of accountability, creating important feedback information to decision-makers and stakeholders regarding the effectiveness of a given program as its implementation proceeds. Feedback, particularly given the future connotation embedded in the theory of sustainability, is a critical aspect of assessment and helps to establish a vision of "correctness" that can be used to modify sustainability strategies. An assessment program that does not encourage feedback or a mechanism to evaluate and make adjustments is unlikely to provide useful information.

As a general procedure, assessment

- Involves the use of empirical data to characterize the critical factors that define the motivating problems
- Is the process of gathering and discussing information from diverse sources in order to form a deep understanding of the salient processes and actors that explain the problem
- Introduces a systematic basis for making inferences about the motivating problem
- Completes the definition, selection, design, collection, analysis, interpretation, and use of information to address the motivating problem
- Encourages a systematic review of the development programs and their objectives

Taken together, these qualities of assessment suggest a simple model that simplifies the procedure into a series of six steps or phases consisting of the following:

1. *Assessment objectives*: This initial phase requires a clear understanding of the intent or focus of assessment. More specifically, this step concentrates on what assessment will observe and measure by establishing the criteria to employ.
2. *Assessment design*: Here the specifics of measurement and data collection are considered. Assessment requires identifying the instrument(s) that will measure the observable elements of the system of interest and the methods used to collect data.
3. *Data collection*: Conducting the assessment is the main activity of this step in the process. Data collection may involve a range of activities from fieldwork, the use of archival sources, and the evaluation of data relevant to the problem.

4. *Analysis*: The data gathered through the assessment process must be translated into information that can guide and support decision-making. Here, a mix of quantitative and qualitative methods may be called upon to provide insight into the patterns and trends revealed by the data.
5. *Documentation and reporting*: Bringing the results to the audience and producing information in a format that is understandable to all concerned is the goal of this step. Communicating the results of assessment in a manner that maximizes the utility and applicability of the findings is the essential goal.
6. *Evaluation and application*: Introducing the results back into the program under assessment is often a missing aspect of the procedure. If the results are not used to improve or redirect the plan or policy, assessment has little value to decision-makers. No matter how small, there is always some action that can be taken to improve or modify a development program.

That model is illustrated in Figure 2.2 and suggests that assessment is primarily a cyclic process. While this model is a useful way to conceptualize the procedural aspects of an assessment program, any approach taken to assess progress toward sustainability relies on a review of data, rubrics, and performance outcomes in relation to the nature of the problem. Without detailed consideration of these three attributes, assessment can be easily frustrated.

Assessing a quality such as environmental sustainability requires assessment rubrics and performance outcomes—the two ingredients in an assessment that may be difficult to derive (Bond et al., 2012). An assessment rubric is a standard of performance that can establish a direction and a means to rate or score progress against a set of evaluative criteria. The assessment rubric is a way to communicate expectations as development strives to achieve "sustainability." Two important questions related to the task of evaluating sustainability have been posed that illustrate the challenge (Bebbington et al., 2007):

1. How can today's operational systems for monitoring and reporting on environmental and social conditions be integrated or extended to provide more useful guidance for those efforts attempting to navigate a transition toward sustainability?
2. How can today's relatively independent activities of research, planning, monitoring, assessment, and decision support be better connected into programs for adaptive management?

With environmental sustainability assessment, the primary focus is to provide decision-makers with an evaluation of an integrated nature–society system over both the short and long term. Assisting those decisions takes the combination of the existing methodologies of environmental impact analysis (EIA) with the evolving practice of sustainability assessment

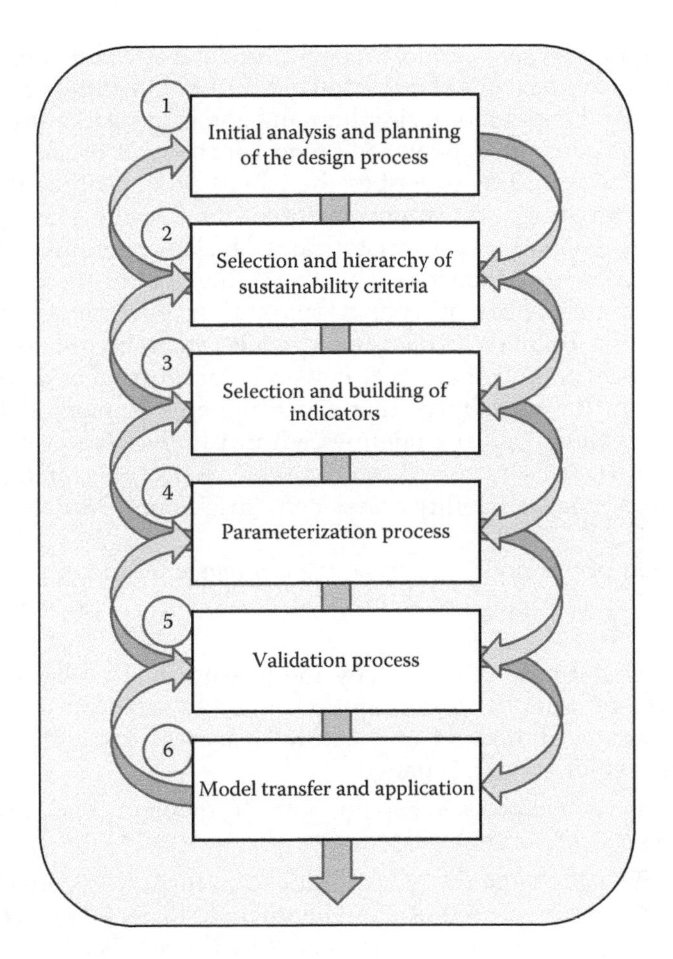

FIGURE 2.2
The sustainability assessment process.

(Gibson, 2006; Kropp and Lein, 2012). Those combined tools form a proactive and pragmatic decision support architecture that can fulfill the desire to comprehend the larger patterns of development and promote environmental sustainability.

2.2 Environmental Impact Analysis

As the entropic influences of human activities became more pronounced and the adverse environmental consequences of those action more widespread, it became obvious that environmental considerations needed to be included early in the decision-making process (Burdge, 1991; Smythe, 1997).

By including previously overlooked environmental attribute damage to the environmental system could be minimized. EIA became the general procedure for reviewing and modifying human actions in relation to both their positive and negative effects on the broad functions of the ecosystem. As a review mechanism, EIA describes the tripart process of identifying the salient characteristics of the environment, evaluating the consequences of human actions on those characteristics, and develop procedures for mitigating those adverse consequences. The overriding goal of these assessment activities is to include careful consideration of the environment in totality into the processes of making plans and policies and incorporate mitigation to protect critical natural or amenity resources for future generations. While the specifics of the procedures that constitute an EIA methodology vary with respect to the technical guidelines, administrative responsibilities, and the legal requirements involved, four fundamental features remain constant that aptly inform sustainability assessment (Erickson, 1994; Lein, 2003):

1. A vision of the environment as the aggregate of things and conditions that surround the living and nonliving components of the ecosystem
2. A decision-centric tool whereby the possible and probable consequences of a human action are examined to avoid an irreversible commitment of amenity and natural resources that will result in adverse environmental change
3. An analysis that seeks to explain both the qualitative and quantitative aspects of the environment
4. An emphasis on the concept of mitigation that directs attention to the investigation of methods and alternatives that may reduce undesirable effects and enhance those identified as beneficial

The connection between EIA and sustainability assessment may be debated (Pope et al., 2004); however, given the need to promote a proactive and long-term view of human development, EIA methods should rest at the foundation of assessment targeting environmental sustainability based primarily on its capacity to examine and identify human–environment interactions and the environmental impacts that may develop.

The concept of an environmental impact may be defined as any alteration of environmental conditions or the creation of a new set of environmental conditions adverse or beneficial, or the creation of a new set of environmental actions, caused or induced by an action or combination of actions. The term "action," initially defined in the context of the National Environmental Policy Act of 1970, describes a wide array of activities undertaken by a decision-making process. Included in the explanation are activities such as the simple granting of a permit, the creation of a program, or plan to the construction of major projects. The significance of this term is that it represents the impetus for change and the potential alterations in the landscape that will produce a new environmental trajectory. This idea is reinforced through the use of the word alteration, a nonspecific term that asks decision-makers to look seriously at the scope of their proposals and examine environmental conditions and their ramifications in their broadest context.

Environmental impacts may also be distinguished by their sequence in the process of inducing change. With respect to sequence, impacts to the environmental system may be characterized as primary or secondary. This distinction is useful when considering how an alteration may transmit through the landscape and present differential effects over time. Primary impacts characterize changed conditions that can be directly attributed to an action. Environmental impacts of this variety are often considered the "first-order" effects that stem from the interaction between the activity and its environmental setting. These changes to the environment are typically immediate and obvious. Secondary impacts are those effects that are indirect or induced by an action or its primary impact. These changes to the environment define "second-order" alterations that can be traced back to the perturbation caused by the action. As a first step in the process of environmental sustainability assessment, EIA encourages decision-makers to

- Develop a complete understanding for the activities or processes active in the landscape
- Gain a detailed understanding of the affected environmental setting
- Project the action or process into the future to determine its larger consequences
- Report the nature of the projected consequences that can be attributed to the activity in question to enable themselves to evaluate their plan, design, or policy and consider alternatives that may be less environmentally problematic

As an assessment tool, the EIA represents a systematic methodology for looking forward and revealing the uncertainties that should inform a decision. This type of review functions to document, predict, and monitor human actions before those actions are taken, and irreversible changes are set into motion. While the procedures that define EIA continue to evolve, its foundations are firmly rooted in a planning process (Lawrence, 2000). This process reflects a unifying first-look analysis that, according to Morgan (1999), (1) applies to all activities that have a significant environmental impact, (2) compares alterative courses of action and possible mitigation measures, (3) results in the documentation of environmental effects that conveys the ramifications of potential unsustainable changes to the environment, (4) encourages broad public participation and review, and (5) incorporates the use of monitoring to gain insight regarding long-term impacts, in order to enable plan and design modifications. The EIA process can be conceptualized as a mental overlay, where an action is superimposed onto its environmental setting. The consequence of that "overlay" sets into motion a type of questioning as pathways of change, modification, and adjustment are examined against the backdrop of exiting environmental conditions. The pragmatic stages that form the analysis of environmental impact have been summarized by Lein (2003):

- *Impact identification*: This initial step defines the task of reading the landscape and exploring the nature and extent of the possible effects an activity may engender. This process can be challenging and relies on scoping and screening activities to collect information on the action, its environment, and the set of environmental conditions that intersect key features of the activity under review.
- *Impact measurement and prediction*: The essential step of looking forward is addressed in this stage of analysis. Estimating effects on the environment in quantitative and qualitative terms concentrates analysis on measuring the magnitude of change and predicting its consequence over time. Measurement and prediction are two of the more problematic aspects of EIA. What to measure, how to measure it, and what measurement explains can frustrate an analysis, but it becomes an essential ingredient of prediction. Prediction of environmental impact can unfold in several ways, including the use of physical models, mathematical or statistical relationships, or through some form of computer simulation. Regardless of the approach taken, the goal of prediction is to obtain basic "what if" insights that can inform the decision.
- *Impact interpretation*: Two distinct operations guide interpretation. First is the question of determining the significance of a given effect (impact). Second has to do with evaluating the magnitude of the change an impact produces. To support decision-making separating magnitude from significance is important. Magnitude is typically arrived at by prediction based on measurement, while significance develops as an expression of the "cost" of the predicted impact to society.
- *Impact communication*: Bringing the results of analysis out into a format that can be understood and shared is the primary goal of communication. Communication centers on the responsibility of disclosure, describing the effects and consequences of an action to inform and support effective decisions.
- *Impact monitoring and mitigation*: Given the persistent nature of the potential changes that may stem from a human activity, the capacity to monitor an action across its life cycle supports the ongoing effort of environmental protection. Modifying actions or plans in response to unwanted or unanticipated changes is critical to the

> goals of sustainability. Several approaches to monitoring have been introduced. Three of the more common are (1) baseline monitoring, which involves the systematic measurement of environmental variables that are used as indicators of landscape conditions, (2) effect monitoring, which looks early on at environmental conditions as an action is implemented, and (3) compliance monitoring, which focuses on the periodic sampling or continuous measurement of environmental variables and their comparison against a set of performance standards.

These elements of EIA suggest a general procedure that concentrates attention on the ramifications of a single plan or action in a defined environmental setting (Figure 2.3). Comprehensive assessment of multiple human developments on the environment has been limited. Cumulative effects, particularly those actions that when viewed in isolation fail to display a significant environmental impact, have important longer-term implications. Recognizing the total effect of separate human decisions on the landscape may be substantially different from a single activity; particularly when they are widely distributed temporally and geographically. Cumulative assessment introduces a broader scale of effects to the analysis, taking into account synergistic and discontinuous patterns that may be introduced into the environmental system. Accounting for these cumulative changes has proved difficult. Recently an approach to assessment has

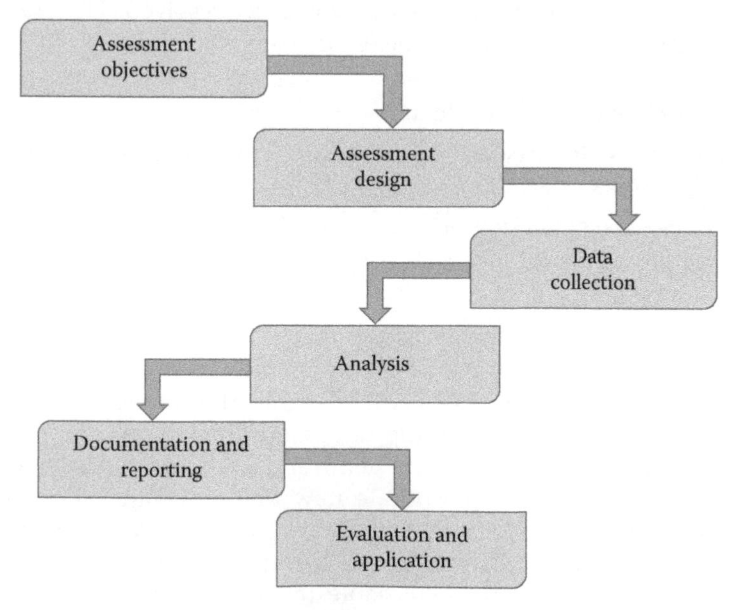

FIGURE 2.3
Generalized stages of environmental assessment.

been suggested that examines the wider implications of human actions. This addition to the general methods of EIA has been referred to as strategic environmental assessment (SEA).

2.3 Strategic Environmental Assessment

The term "strategic environmental assessment" was introduced in the late 1980s and continues to evolve conceptually and procedurally (Tetlow and Hanusch, 2012). Initially, SEA developed out of the need to move beyond the single action or project-based focus of EIA. Presently, SEA represents a mechanism to identify and evaluate the potential impacts of policies, plans, and programs. The primary goal of SEA is to arrive at long-term sustainable solutions (Pope et al., 2005; Pope and Grace, 2006). While the instruments that form SEA vary, the term is used largely to describe an environmental assessment process for policies, plans, or programs that are approved much earlier than the authorizations needed for individual projects (Lee and Walsh, 1992).

Justification for conducting an SEA grew from the need to take into consideration the potential environmental effects of activities that do not conform to the typical definition of a project. Therefore, a plan or policy can have an environmental impact, even though no physical construction on the land is involved. The strategies of development, examined early in the planning process, can be critically examined and key issues resolved. The SEA, therefore, overcomes certain EIA deficiencies such as

1. A restrictive project-level focus
2. An inability to account for impacts that result from ancillary developments
3. A foreclosure of alternatives
4. A failure to address cumulative effects

SEA provides a means by which policy instruments are developed based on a broader set of perspectives, objectives, and constraints beyond those normally considered in an EIA. According to this model, SEA requires decision-makers to adopt a holistic understanding of the environmental and social implications of proposed plans and policies. This broader scope enhances the potential to incorporate new objectives and constraints in plan formulation, the substitution of alternative objectives, policy implementation, and conflict resolution, all designed to move the decision-making arena toward more sustainable outcomes (Brown and Therivel, 2000). Encouraging sustainable decisions, using SEA as the vehicle, concentrated efforts on setting environmental quality goals, strengthening

institutions to attain those goals, and improving procedures and methods of long-term assessment. Through these enhancements, SEA creates a systematic, integrated framework to incorporate specific sustainability principles early on in the plan or policy deliberation. Additionally, the early examination of environmental impact requirement of SEA directs analysis at a much higher level and at a more comprehensive scope than typically accomplished with EIA. In this regard, SEA can facilitate the detection, management, and monitoring of activities likely to fall under the heading of a cumulative impact, forcing examination to explore a large geographic footprint across long time horizons.

An important property of SEA is its capacity to function with a tier system of analysis that complements both EIA and programs' aims at sustainability assessment. Tiering places SEA on the continuum of the assessment spectrum with an obvious focus on a more strategic view of the environment relative to the motivating needs that become the subject of plans, policies, or programs. Procedurally, SEA gives attention to the nature of decisions, their intensions, orientations, and methods of guidance. Any decision suggests a strategy, and as such its intent can be overlaid onto its environmental setting, reviewed and replaced, which can add continuity to decisions by capitalizing on the realization that any policy, plan, or program establishes a direction of change. That change sets into motion a potentially continuous stream of activities within a given environmental setting. To evaluate that stream of change, SEA requires flexibility and adaptiveness as the range of possible alterative choices surrounding a decision are explored and the forces acting on the decision are integrated and reconciled within an environmental context.

Several attempts have been made to outline the SEA process following a logic and language similar to EIA (Fischer, 2010; Therivel and Paridario, 2013). Although not a linear methodology, there are five critical stages that support the process (Figure 2.4):

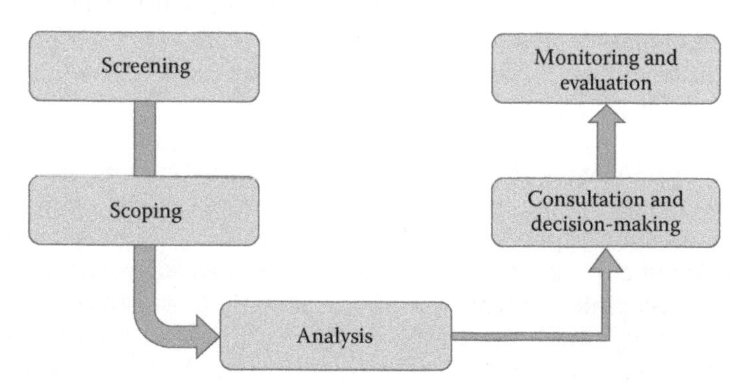

FIGURE 2.4
Generalized stages of SEA.

1. *Screening*: Screening determines whether an SEA is required or not. This determination centers on the potential for the policy, plan, or program to have a significant effect on the environment.
2. *Scoping*: Scoping reduces the range of possible effects to those considered to pose the greatest impact and therefore become the set of effects that assessment will concentrate on. The main considerations in the scoping process are the evaluation of significant effects on the environment and the examination of alternatives.
3. *Analysis*: Analysis documents the potential environmental effects. Through analysis information is gathered relative to the critical issues that impact environmental sustainability. Considerations such as biodiversity loss, population development, human health, fauna, flora, soils, air and water quality, climate change, material assets, and cultural heritage suggest the list of factors examined. Examination also includes detailed consideration of secondary, cumulative, and synergistic effects.
4. *Consultation and decision-making*: At the conclusion of analysis, the documentation of environment effects is reviewed. Review of the findings informs both the public and decision-makers. Recommendations are then employed to rethink motivating agendas. Consultation, therefore, encourages dialogue and is often a useful learning process for all parties involved in the decision. Ideally, decision-makers are shown the options to their plan, the likely effects it will have, the available choices, and the consequence of a poor (ill-informed) decision.
5. *Monitoring and evaluation*: Monitoring the extent to which the environmental objectives and recommendations made in the SEA are met is an essential step to insure that environmental protection is maintained within the policy arena. Checking on progress, evaluating cumulative effects, and gaining vital early warning information of regional-scale changes in relation to the region's natural assets guide the rational implementation of plans. Monitoring also provides feedback information that supports formal evaluation, renewal, or revisions of the policy or plan.

These steps that suggest the SEA process carry sustainability assessment to the next logical step in examining human impact of the environmental system. As human impact is largely the consequence of human decisions that develop out of policy choices, the SEA becomes a successful instrument when it can integrate environmental considerations as polices are taking shape. The missing piece within this framework, however, is the answer to the question of whether a project, plan, or policy initiative is actually sustainable. The assessment of sustainability introduces a unique set of challenges that expands the role of environmental assessment as a decision instrument.

2.4 Environmental Sustainability Assessment

Assessing environmental sustainability is motivated by the desire to direct development toward outcomes that reduce cross-generation declines in critical ecological services. The purpose of assessment is to provide decision-makers with an evaluation of a connected nature–society system and to offer guidance as to which actions society can likely sustain (Ravetz, 2000; Ness et al., 2007; Singh et al., 2012). Through this evaluation, the goal is to assure decision-makers that plans and activities undertaken make an "optimal" contribution to forms of development that achieve a level of fit within their environmental setting. The concept of fit implies the capacity for activities to function across an extended time horizon in a manner balanced with the capabilities of the ecological resource base, provided the needed support is available within the environmental setting. Environmental sustainability becomes a "state of being" that defines a set of conditions that can be identified according to selected criteria (Pope et al., 2004). A sustainable environmental system, therefore, has a quality that proposed human activities should preserve. Explaining that quality remains an aggressively debated topic, nevertheless, accepting the premise that a sustainable environmental system is characterized by a set of conditions and trends that can continue "indefinitely," assessment can be conducted based on the need to answer five framing questions (Lein, 2014):

1. What is the present state of the environmental system?
2. Is that state considered sustainable?
3. What actions/processes are driving the system forward?
4. Are there indications that the environmental system is degrading?
5. Can that information be incorporated into policy decisions to guide the future?

Addressing these questions hinges on the formulation and application of tractable indices that can be organized into a well-defined method of analysis.

2.5 Assessment Indicators

An index is a summary measure that attempts to communicate a salient characteristic of a system or situation of interest. Measurement of an index provides relevant insight into the behavior or status of a system that (1) quickly informs a target audience, (2) reduces complexity of a given condition, and (3) clearly identifies qualities or trends. An indicator useful to the purpose of environmental sustainability assessment shares one or more common attributes (Bauler, 2012; Dahl, 2012; Hák et al., 2012):

- General relevance: Consideration of relevance determines the degree to which an indicator characterizes the environment and facilitates the evaluation of process and change.
- Conceptual integrity: This attribute supports and justifies a rationale, which may be theoretical or pragmatic, for using the indicator.
- Reliability: This refers to focusing on the question of how effective the indicator is and the level of explanation it provides.
- Scale appropriateness: Scale expresses the capacity of the indicator to detect the desired environmental condition at the appropriate spatial and temporal resolution.
- Statistical sensitivity: This relates to the degree of measurement precision and accuracy that can be derived from the indicator as well as the degree of confidence that surrounds its application.
- Robustness: This pertains to the ability of the indicator to yield consistent results under a range of conditions.

Sustainability assessment has encouraged development of a wide array of indicators. These indices are as diverse and contested as the concept of sustainability (Mayer, 2008; Lein, 2014). When assessing environmental sustainability, indicators must be capable of determining (1) where ecological functioning and human activities intersect with pronounced intensity now and into the future and (2) where maintaining balance between ecological functions and human activity is critical to resolving conflicts in those locations where existing development trends induce potentially adverse environmental patterns that can be extrapolated into the future (Mascarenhas et al., 2010; Moldan et al., 2012).

Arriving at the optimal spatial scale where these conditions can be understood is among the most difficult aspects of examining environmental sustainability. The critical decision of scale relates to the relative or absolute spatial dimension or expression that defines an object of quality. In the context of assessment, scale has two important explanations: analysis scale and phenomena scale. Analysis scale speaks to the level of data aggregation or resolution required to observe (measure) a given problem or situation. That detail influences how a condition can be studied and the level of generalization and error that might be induced. Typically, analysis scale describes the physical area that is represented by the data, characterizing its regional context whether that is defined in terms of an urban administrative district, watershed, or a larger spatial unit. Phenomena scale refers to the size at which specific human or physical processes manifest. Processes may be small-scale (micro) activities or extend across the spectrum to global scales. This definition defines the "true" expression of the patterns, processes, or objects under investigation and often those features are inherently scale dependent. The targets of environmental sustainability exemplify the range of these traits that pose challenges to measurement and meaning. Adopting a regional or meso-scale level of analysis has been suggested as the most appropriate level of resolution for the purposes of environmental sustainability assessment (Graymore et al., 2008, 2010). At this level of detail, the complex interactions that exist between ecological, social, and economic phenomena can be more closely integrated with a landscape unit that can be delineated on the basis of well-understood anthropogenic criteria (such as the watershed). Operating at the watershed scale, analysis can be supported by well-defined indicators whose variability in the landscape can be related to development pressures. The landscape subject to human drivers displays important shifts in the sustainability of the environmental system and is signaled by deviations in the indicator set. For example, the removal of a biological community, modification of a habitat, or alteration in the size and configuration of a land use parcel may serve to evidence unsustainable development and function to degrade ecological stability over time. By focusing attention to this regional scale, trends in development—expressed as land use and land cover change—contribute to the variations in the degree of naturalness within the watershed unit. The physical replacement of land cover, coupled with measureable geometric patterns that theory links to known disturbances in energy flows, resource availability, and biodiversity, should encourage careful reconsideration of existing programs and guide management actions toward development plans that avoid those disturbances and produce patterns within the human landscape that are, by definition, sustainable.

Realizing environmental sustainability at the regional scale of analysis concentrates attention on the creation and implementation of tractable methodologies (Moldan et al., 2012). Those methods are critical and share a common goal: to encourage well-informed and timely reactions to changing landscape conditions. Several key considerations in the development of an assessment method have been presented by Roberts (2006) and include

- A definition of development that identifies the key issues to be considered
- A determination of the range of interactions between the environmental, social, and economic dimension of development
- A determination of the scope and structure of analysis
- A selection of the appropriate indicators that can be applied to monitor and measure progress and serve as a baseline for future evaluation

2.6 Sustainability Indicators

Indicator science has had a long history in landscape analysis (Chikishev, 1973). The range of measures and indices developed and presented in the literature is extensive and continues to evolve as new theories are tested and new approaches are placed into application. Indicators designed to support the assessment and monitoring of environmental sustainability display an equally diverse range that challenges both indicator selection and the data required to populate the calculations used to derive a measure (Heink and Kowarik, 2010; Huang et al., 2015). When directed at the question of long-term environmental sustainability, indicators are analytical and interpretive tools that express the ecological dynamics of the regional landscape (Venturelli and Galli, 2006), and although the term is often used ambiguously, an indicator is a measure or component from which conclusions on the phenomenon of interest can be inferred (Heink and Kowarik, 2010). The essential quality of any sustainability indicator is its capacity to provide a signal of performance in relation to the planning directives in place designed to guide development. There are three critical elements to this task (Alberti, 1996):

1. Key variables that describe the coupled urban/environmental system and their interrelationship
2. Measureable objectives and criteria that facilitate assessment
3. Feedbacks that send signals of performance to decision makers

The decision of which variables to measure and which criteria to use depends largely on how sustainability is defined. At present, there is little guidance to help make that choice which has contributed to a confusing mix of sustainability metrics that function across a range of spatial scales (Moldan et al., 2012). A review of sustainability indicators presented by Mori and Christodoulou (2012) focused on urban sustainability contends that no major index or indicator has been introduced that satisfies all aspects of the sustainability concept. A more considered view of the role of indicators in the assessment process is given by Mascarenhas et al. (2010). Rather than attempting to fulfill requirements implied by definitions of sustainability, indicators are more effective if appropriately linked to specific strategies that have been implemented to achieve a desired outcome. Given the realities that sustainability is not easily defined nor measured, indicators serve as useful surrogates that capture one or more aspects of a coupled human/environmental system and can therefore reflect the status of that system in a selective manner.

While the literature abounds with indicators, there is no consistent typology that clearly brings these measures into an organized framework. Sustainability indicators are often categorized as global, national, local, or with respect to social, economic, or ecological themes (Layke et al., 2012). Several of the more common include (after Alberti, 1996)

- Indicators of urban pattern, which include variables such as population density, land cover, derelict land areas, and urban mobility
- Indicators of urban flows, which identify variables such as water consumption, wastewater generation, energy consumption, solid waste and production waste recycling, and waste disposal
- Indicators of environmental quality, which focus on measures of water quality, air quality, noise levels, proximity to green space, and habitat quality

Other indicators cited in the literature include a range of healthy city variables that incorporate measures of mortality, births, health services, living space, poverty, and employment (Table 2.1), as well as a set of urban sustainability indicators reported in van Dijk and Mingshun (2005) that index urban status, urban coordination, and urban potential. Other factors commonly applied to assess environmental sustainability include a selection of landscape metrics and a suite of composite indices such as the Ecological Footprint, Green City Index, Environmental Performance/Environmental Sustainability Index, Human Development Index, and Happy Planet Index (Wu and Wu, 2012; Huang et al., 2015).

The concept of a landscape metric refers exclusively to indicators developed from categorical map patterns. Metrics, taken from the spatial arrangement of thematic data, are essentially algorithms that quantify geographic characteristics of pattern, focusing on either the composition of the map without reference to spatial attributes or the spatial configuration of the map based on the specific arrangements and juxtaposition of map categories. As fundamental descriptions of landscape pattern, the relationship between structure and function strongly influences ecological processes, biotic abundance, and diversity. From an analytical perspective, changes in landscape structure alter landscape function and vice versa. Three important levels of metric quantification can be noted:

1. *Patch-level metrics*—defined for individual patches and characterize the spatial character and context of landscape patches.
2. *Class-level metrics*—integrated over all the patches of a given type or category.
3. *Landscape-level metrics*—integrated over all patch types or categories across the full extent of the landscape (scene). Landscape-level metrics can be subdivided into eight measurement groupings:
    a. Area/density/edge metrics
    b. Shape metrics
    c. Core-area metrics

TABLE 2.1
Selected Sustainability Indicators

Objective	**Environmental health**		
Policy categories	**Environmental burden of disease**	**Water (effects on humans)**	**Air pollution (Effects on humans)**
Indicators	1. Environmental Burden of disease	2. Adequate sanitation 3. Drinking water	4. Indoor air pollution 5. Urban particulates 6. Local ozone
Objective	**Ecosystem vitality**		
Policy categories	**Air pollution (effects on ecosystems)**	**Water**	**Biodiversity and habitat**
Indicators	7. Regional ozone 8. Sulfur dioxide emissions	9. Water Quality Index 10. Water stress	11. Conservation Risk Index 12. Effective conservation 13. Critical habitat protection 14. Marine protected areas
Policy categories	**Productive natural resources**	**Productive natural resources**	**Productive natural resources**
Policy subcategory	**Forestry**	**Fisheries**	**Agriculture**
Indicators	15. Growing stock	16. Marine Trophic Index 17. Trawling intensity	18. Irrigation stress 19. Agricultural subsidies 20. Intensive cropland 21. Burnt land area 22. Pesticide regulation
Policy categories	**Climate change**		
Indicators	23. Emissions per capita 24. Emissions per electricity generated 25. Industrial carbon intensity		

d. Isolation/proximity metrics
e. Contrast metrics
f. Contagion/interspersion metrics
g. Connectivity metrics
h. Diversity metrics

These metrics are carefully reviewed and extensively described by MacGarigal and Marks (1995). A suite of improved landscape metrics that are independent of the characteristic variations has been introduced by Frohn (1997). A core set of landscape metrics useful for landscape analysis and environmental sustainability assessment selection includes the following:

- *Diversity*—expressed according to

$$H = -\text{sum}\,(p \times \ln(p))$$

  where p is the proportion of each class in the kernel.
- *Dominance*—a metric calculated according to

$$D = H_{max} - H$$

  where
  - H is a measure of diversity
  - H_{max} is the maximum diversity based on the natural logarithm of the number of different classes present in the kernel

- *Fragmentation*—expressed according to

$$F = \frac{n-1}{c-1}$$

  where n represents the number of classes present in the kernel and c the number of cells directing the calculation (9, 25, or 49).
- *Relative richness*—determined by

$$R = \frac{n}{n_{max}} \times 100$$

  where
  - n is the number of different classes present in the kernel
  - n_{max} is the maximum number of classes in the scene

- *Fractal dimension*—computed over the entire surface using a moving window operation typically set to a 3 row by 3 column design; the fractal dimension is found from the formula

Line

= log N^1/log N
= 1 * log N/log N
D = 1

Cube

= log N^3/log N
= 3 * log N/log N
D = 3

Square

= log N^2/log N
= 2 * log N/log N
D = 2

Composite indices are calculated from a mathematical formulation of specific variables that are functionally related and can be aggregated into a single score. The ecological footprint is an area-based score related to carrying capacity concepts that expresses the land area required to provide the energy and material resources consumed and the wastes discharged as a function of population. The Green City Index is a composite of 30 indicators and gives a regional accounting of relative environmental performance. The Environmental Performance Index measures the degree to which regions meet established targets and can be used to compare regions and trends over time. The Human Development Index assesses the level of human and social development and takes into account life expectancy, education, and standard of living. The Happy Planet Index is an alternative to the Human Development Index and combined human welfare variables with variables that explain human consumption of natural resources. The Sustainable Society Index is a highly aggregated formulation of 23 well-being dimensions, 7 categories, and 21 indicators.

An indicator can be a powerful means of expressing important dimensions of the environment and enabling a clear understanding of complex relationships that can guide management decision-making. For an issue as challenging as environmental sustainability, adequate indicators are essential; however, considerable work remains to be done in their refinement and consistency as they are applied to examine progress toward a sustainable future. Several screening questions can be posed to assist with the selection of a relevant indicator (Table 2.2). Answers to these questions provide a basis to examine how well the index captures aspects of policy

TABLE 2.2

Sustainable Community Indicator Checklist

Does the indicator address the carrying capacity of the natural resources—renewable and nonrenewable, local and nonlocal—that the community relies on?
Does the indicator address the carrying capacity of the ecosystem services upon which the community relies, whether local, global, or from distant sources?
Does the indicator address the carrying capacity of aesthetic qualities—the beauty and life-affirming qualities of nature—that are important to the community?
Does the indicator address the carrying capacity of the community's human capital—the skills, abilities, health, and education of people in the community?
Does the indicator address the carrying capacity of a community's social capital—the connections between people in a community?
Does the indicator address the carrying capacity of a community's built capital—the human-made materials (buildings, parks, playgrounds, infrastructure, and information) that are needed for quality of life and the community's ability to maintain and enhance those materials with existing resources?
Does the indicator provide a long-term view of the community?
Does the indicator address the issue of economic, social, or biological diversity in the community?

whose outcomes are realized in the future. The context of the future is an essential feature of the assessment problem and presents one of the more critical and easily overlooked questions when exploring decisions deemed to be environmentally sustainable. The unique issues that surround the future and the specific requirement to include the future as the central element of environmental sustainability assessment are explored in the chapters to follow.

2.7 Summary

The capacity to stand back and examine how a plan or agenda is working is an essential part of the environmental planning and management process. Assessment is the essential step in gaining the insight required to evaluate progress in achieving the goals and objectives stated in strategies designed to maintain a sustainable environmental system. Environmental assessment provides the first step in the larger understanding of sustainability and offers both theory and method that can be adapted to incorporate the requirements of an assessment procedure targeting the sustainability question. Assessment, however, implies measurement, and while the precise elements of the environmental system that lend themselves to the task of explaining sustainability as a single variable remain elusive, a range of indicators can be employed that summarize critical signals of the environmental system that explain environmental process, describe patterns and relationships, and can therefore be employed as surrogate measures that inform decision-making.

References

Alberti, M. (1996). Measuring urban sustainability. *Environmental Impact Assessment Review, 16*(4), 381–424.

Bauler, T. (2012). An analytical framework to discuss the usability of (environmental) indicators for policy. *Ecological Indicators, 17*, 38–45.

Bebbington, J., Brown, J., and Frame, B. (2007). Accounting technologies and sustainability assessment models. *Ecological Economics, 61*(2), 224–236.

Bond, A., Morrison-Saunders, A., and Pope, J. (2012). Sustainability assessment: The state of the art. *Impact Assessment and Project Appraisal, 30*(1), 53–62.

Bond, A. J. and Morrison-Saunders, A. (2011). Re-evaluating sustainability assessment: Aligning the vision and the practice. *Environmental Impact Assessment Review, 31*(1), 1–7.

Brown, A. L. and Thérivel, R. (2000). Principles to guide the development of strategic environmental assessment methodology. *Impact Assessment and Project Appraisal, 18*(3), 183–189.

Burdge, R. J. (1991). A brief history and major trends in the field of impact assessment. *Impact Assessment, 9*(4), 93–104.

Chikishev, A. (1973). *Landscape Indicators: New Techniques in Geology and Geography.* Consultants Bureau, New York.

Dahl, A. L. (2012). Achievements and gaps in indicators for sustainability. *Ecological Indicators, 17,* 14–19.

Erickson, P. A. (1994). *A Practical Guide to Environmental Impact Assessment.* Academic Press Inc, Cambridge, MA.

Fischer, T. B. (2010). *The Theory and Practice of Strategic Environmental Assessment: Towards a More Systematic Approach.* Routledge.

Frohn, R. C. (1997). *Remote Sensing for Landscape Ecology: New Metric Indicators for Monitoring, Modeling, and Assessment of Ecosystems.* CRC Press, Boca Raton, FL.

Gibson, B., Hassan, S., and Tansey, J. (2013). *Sustainability Assessment: Criteria and Processes.* Routledge, New York.

Gibson, R. B. (2006). Sustainability assessment: Basic components of a practical approach. *Impact Assessment and Project Appraisal, 24*(3), 170–182.

Glasson, J., Therivel, R., and Chadwick, A. (2013). *Introduction to Environmental Impact Assessment.* Routledge, New York.

Graymore, M. L., Sipe, N. G., and Rickson, R. E. (2008). Regional sustainability: How useful are current tools of sustainability assessment at the regional scale? *Ecological Economics, 67*(3), 362–372.

Graymore, M. L., Sipe, N. G., and Rickson, R. E. (2010). Sustaining human carrying capacity: A tool for regional sustainability assessment. *Ecological Economics, 69*(3), 459–468.

Hák, T., Moldan, B., and Dahl, A. L. (Eds.) (2012). *Sustainability Indicators: A Scientific Assessment,* Vol. 67. Island Press, Washington, DC.

Heink, U. and Kowarik, I. (2010). What are indicators? On the definition of indicators in ecology and environmental planning. *Ecological Indicators, 10*(3), 584–593.

Huang, L., Wu, J., and Yan, L. (2015). Defining and measuring urban sustainability: A review of indicators. *Landscape Ecology, 30*(7), 1175–1193.

Jay, S., Jones, C., Slinn, P., and Wood, C. (2007). Environmental impact assessment: Retrospect and prospect. *Environmental Impact Assessment Review, 27*(4), 287–300.

Kropp, W. W. and Lein, J. K. (2012). Assessing the geographic expression of urban sustainability: A scenario based approach incorporating spatial multicriteria decision analysis. *Sustainability, 4*(9), 2348–2365.

Lawrence, D. P. (2000). Planning theories and environmental impact assessment. *Environmental Impact Assessment Review, 20*(6), 607–625.

Layke, C., Mapendembe, A., Brown, C., Walpole, M., and Winn, J. (2012). Indicators from the global and sub-global Millennium Ecosystem Assessments: An analysis and next steps. *Ecological Indicators, 17,* 77–87.

Lee, N. and Walsh, F. (1992). Strategic environmental assessment: An overview. *Project Appraisal, 7*(3), 126–136.

Lein, J. K. (2003). *Integrated Environmental Planning: A Landscape Synthesis.* John Wiley & Sons, New York.

Lein, J. K. (2014). Toward a remote sensing solution for regional sustainability assessment and monitoring. *Sustainability, 6*(4), 2067–2086.

MacGarigal, K. and Marks, B. J. (1995). *FRAGSTATS: Spatial Pattern Analysis Program for Quantifying Landscape Structure. USDA Forest Service.* Pacific Northwest Research Station, Portland.

Mascarenhas, A., Coelho, P., Subtil, E., and Ramos, T. B. (2010). The role of common local indicators in regional sustainability assessment. *Ecological Indicators, 10*(3), 646–656.

Mayer, A. L. (2008). Strengths and weaknesses of common sustainability indices for multidimensional systems. *Environment International, 34*(2), 277–291.

Moldan, B., Janoušková, S., and Hák, T. (2012). How to understand and measure environmental sustainability: Indicators and targets. *Ecological Indicators, 17*, 4–13.

Morgan, R. K. (1999). *Environmental Impact Assessment: A Methodological Approach*. Springer Science and Business Media, New York.

Mori, K. and Christodoulou, A. (2012). Review of sustainability indices and indicators: Towards a new City Sustainability Index (CSI). *Environmental Impact Assessment Review, 32*(1), 94–106.

Ness, B., Urbel-Piirsalu, E., Anderberg, S., and Olsson, L. (2007). Categorising tools for sustainability assessment. *Ecological Economics, 60*(3), 498–508.

Pope, J., Annandale, D., and Morrison-Saunders, A. (2004). Conceptualising sustainability assessment. *Environmental Impact Assessment Review, 24*(6), 595–616.

Pope, J. and Grace, W. (2006). Sustainability assessment in context: Issues of process, policy and governance. *Journal of Environmental Assessment Policy and Management, 08*(03), 373–398.

Pope, J., Morrison-Saunders, A., and Annandale, D. (2005). Applying sustainability assessment models. *Impact Assessment and Project Appraisal, 23*(4), 293–302.

Ravetz, J. (2000). Integrated assessment for sustainability appraisal in cities and regions. *Environmental Impact Assessment Review, 20*(1), 31–64.

Roberts, P. (2006). Evaluating regional sustainable development: Approaches, methods and the politics of analysis. *Journal of Environmental Planning and Management, 49*(4), 515–532.

Singh, R. K., Murty, H. R., Gupta, S. K., and Dikshit, A. K. (2012). An overview of sustainability assessment methodologies. *Ecological Indicators, 15*(1), 281–299.

Smythe, R. B. (1997). The historical roots of NEPA. In: Clark, E. R. and Canter, L. W. (Eds.). *Environmental policy and NEPA: Past, present, and future* (pp. 3–14). CRC Press, Boca Raton, FL.

Tetlow, M. and Hanusch, M. (2012). Strategic environmental assessment: The state of the art. *Impact Assessment and Project Appraisal, 30*(1), 15–24.

Therivel, R. and Paridario, M. R. (2013). *The Practice of Strategic Environmental Assessment*. Routledge, New York.

van Dijk, M. P. and Mingshun, Z. (2005). Sustainability indices as a tool for urban managers, evidence from four medium-sized Chinese cities. *Environmental Impact Assessment Review, 25*(6), 667–688.

Venturelli, R. C. and Galli, A. (2006). Integrated indicators in environmental planning: Methodological considerations and applications. *Ecological Indicators, 6*(1), 228–237.

Wu, J. and Wu, T. (2012). *Sustainability Indicators and Indices: An Overview. Handbook of Sustainable Management* (pp. 65–86). Imperial College Press, London.

3

Prediction, Uncertainty, and Environmental Sustainability

Environmental sustainability is a destination that is reached at some point in the future. Efforts to both define environmental sustainability and develop strategies to assess its status carry the implicit assumption that action or decisions made in the present can be extrapolated into the future and their outcomes understood. The future, however, does not give up its secrets willingly, and any attempt to assess how a policy, plan, or program will unfold over time must contend with the presence of uncertainty and recognize the limitations inherent to the art and science of prediction. In this chapter, the nature of uncertainty, its connection to prediction, and how it influences sustainability assessment and decision-making are examined.

3.1 Uncertainty and Time

Any discussion of environmental sustainability inevitably concerns time. With a focus on the future and the need to track progress over extended time horizons, sustainability planning must confront effective problems of choice and action with an ability to (1) decipher how change in a coupled human–environment system takes place and (2) determine what the impact of those changes may imply. Answers to both questions are shrouded in uncertainty due largely to the realization that the environment is a complex and dynamic system involving interactions that are often poorly understood. The concept of uncertainty has received considerable attention in environmental decision-making (Milliken, 1987; Lemons, 1997; Abbott, 2005). While a concise definition is challenging to offer, uncertainty in the context of sustainability assessment and planning represents the gap between what is known and what needs to be known to make a correct decision. Uncertainty can be considered a type of ignorance that stems from a perceived lack of knowledge by an individual or group (Abbott, 2005). That uncertainty imparts an influence relative to the decisions or actions being undertaken, although the nature of that influence may not be recognized. Perhaps a more direct definition explains uncertainty as a perceived inability to predict something accurately. This particular expression of the term develops from the perception that there is insufficient information regarding cause and effect relationships and that this lack of information frustrates both the ability to explain the likelihood of future events and to predict accurately the outcome of a decision (Milliken, 1987). In isolation, these ideas couch uncertainty in very general terms; however, there are dimensions of uncertainty that are extremely relevant to the question of environmental sustainability and its evaluation. These additional characteristics include three categories of uncertainty identified by Heijden (1996): (1) risks, where the probabilities of a given outcome are known; (2) structural uncertainties, describing possible events where known causal relationships give an indication of likelihood; and (3) knowables, where ignorance permeates all aspects of the problem. Uncertainty may also be subdivided into four dimensions that focus on the connection between the present and the future state of the environmental system (Abbott, 2005). The dimensions of uncertainty that influence sustainability decision-making are as follows (Figure 3.1):

- *Causal uncertainty*—defining the lack of knowledge regarding basic process and relationships (physical, social, economic)
- *Human and organizational uncertainty*—explaining the ignorance characterizing the actions and future intentions of people and organizations
- *External uncertainty*—characterizing the wider social, physical, and economic environment and how each relates to the influences of a given outcome
- *Chance uncertainty*—recognizing the presence of randomness and truly unknowable events that impact a given outcome

There is also a process-related dimension to the nature of uncertainty (Bodansky, 1991; Underwood, 1997; Walker et al., 2013; Wu et al., 2006). This aspect tends to reveal a type of ignorance concerning the appropriateness of

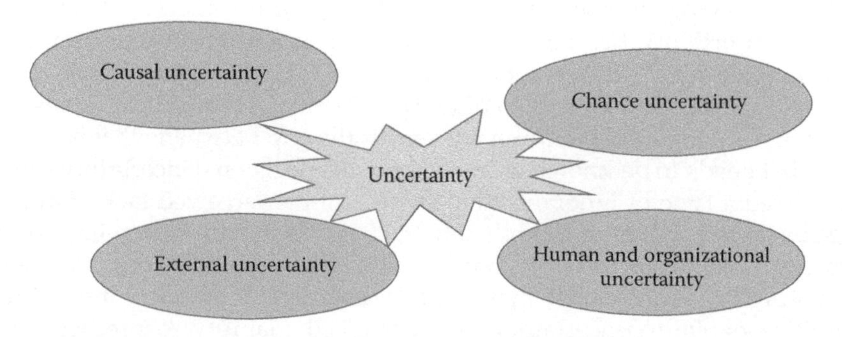

FIGURE 3.1
The dimensions of uncertainty.

value judgments and the aspirations of individuals and groups involved in or impacted by decision-making.

Developing an agenda to achieve environmental sustainability focuses on changing the expected future by altering or modifying social, economic, and environmental connections assumed to conflict between the present and an uncertain future state. Deriving that agenda and realigning connections toward the desired sustainable state centers not simply on the problem of uncertainty, but also recognizes the sources of uncertainty that may frustrate long-term goals. Uncertainty is introduced into decision-making from four principal sources. First, decision-making must contend with the uncertainty related to the data and information used to guide and evaluate choice. Data are often highly variable with respect to precision and accuracy. In other situations, data may be incomplete, old, or miss-specified. Second, uncertainty may be introduced through the analytic methods used to support decision-making. Uncertainty related to method may occur as a consequence of poor conceptual understanding of the problem or from errors introduced through the analytic operations and models used. Natural stochasticity is a third source of uncertainty that highlights the complexity and indeterminacy of the environmental system involved. Natural systems display a degree of randomness that conspires to limit their absolute predictability. Stochastic systems are approached in terms of probabilities that attempt to reduce uncertainty to a more manageable level. Finally, uncertainty can be introduced through the subject-technical judgments and value statements made during all stages, from problem identification analysis, choice, to implementation.

Understanding the nature of uncertainty and the interrelationships between those aspects that pertain to the external environment, chance, and the conceptualization of causality taken together with those related to the decision process reduces uncertainty to a more manageable status (Reckhow, 1994). However, for many situations that surround sustainability planning and assessment, appropriate or accepted theories and relevant models are not available. Decision-making under these circumstances, therefore, is characterized by what has been referred to as deep uncertainty (Lempert et al., 2003). Deep uncertainty assumes a position on the continuum just below ignorance (Faucheux and Froger, 1995). Here, as is common with the concept of sustainability planning, the decision problem has no historical precedent. Therefore, there is no concrete foundation on which to form a complete understanding of the consequences associated with a decision. Furthermore, there is an element of reversibility related to the decision and the changes it may induce within the environmental system. That obvious potential for change adds additional uncertainty to the decision since the set of options available can change through time as a consequence of the multidimensional interactions that may form between the social, economic, and environmental components that can be anticipated to propel the system toward a sustainable pattern.

In light of its significance, managing uncertainty and evaluating its effect on the decision and the decision environment become paramount. Analytic approaches for identifying and assessing uncertainty require direct answers to the following questions that frame the evaluation and ask for a critical examination of the factors surrounding the decision:

- What are the primary sources of uncertainty?
- How large are the uncertainties?
- Which uncertainties are the most important?
- How does uncertainty impact the decision?

Means to identify and explain uncertainty include approaches to determine the confidence bounds for data and model parameters, estimate probabilities, conduct sensitivity analyses, and develop alternative models (Figure 3.2). Strategies to act in the face of uncertainty adopt fairly pragmatic solutions and follow one of four decision styles:

1. *Ignore uncertainty*: This strategy generally involves making a decision and waiting on its outcome. It also requires decision-makers to accept a large degree of risk/uncertainty with respect to both the decision and its eventual outcome.
2. *Delay the decision*: By delaying the decision, the assumption is that uncertainty may decline with time. However, while some uncertainty may disappear, new sources may materialize.
3. *Reduce uncertainty*: Through acquisition of information or by means of sensitivity analysis, some aspects of uncertainty may be better understood and controlled for.
4. *Accept uncertainty*: A position that directs consideration to the precautionary principle and its role in moving forward with a decision given the uncertainty inherent to the situation.

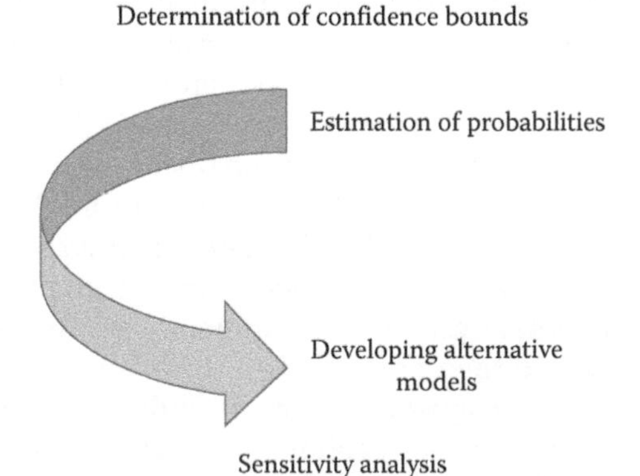

FIGURE 3.2
Steps involved in the management of environmental uncertainty.

3.2 Precaution and Environmental Sustainability

Every decision takes place within an atmosphere punctuated by elements that are unknown. Although the degree of those unknowns may differ, the presence of uncertainty suggests that behind any decision rests an element of risk. Embarking on a course of action assumed to produce a sustainable outcome carries a unique element of risk simply because sustainability is essentially a prediction based on the extrapolation of current knowledge into the future. The level of uncertainty this introduces is common to many issues related to environmental decision-making. In response to decision risk, policy-makers have relied on the precautionary principle as one way to contend with the unknowns that surround a decision.

The precautionary principle has been widely explored in environmental planning and management (Deville and Harding, 1997; Kriebel et al., 2001; O'Riordan, 1994). While numerous definitions exist, the concept implies the need to exercise caution in advance of a decision. The precautionary principle defines a type of informed prudence that encourages decision-makers to anticipate the potential negative aspects of a choice in proportion to the risks and the feasibility of the proposed action under consideration (Gullett, 1997). In the broad field of environmental management, the precautionary principle has been employed to direct rational decision-making by introducing careful consideration for the irreversibility of possible future consequences. That concern for irreversibility finds useful application when considering environmental sustainability. Adopting a precautionary approach represents a shift from the historically accepted policy-making paradigm that relies on the infallibility and comprehensiveness of scientific knowledge (Peel, 2005). The current belief that environmentally sustainable development can be achieved is instructive in this regard. Assessing sustainability based on the application of targets or preset thresholds for key environmental indicators suggests a nearly intractable level of uncertainty surrounding the correctness of any policy decision. Uncertainty, however, does not render sustainability meaningless; rather, it tends to reinforce target-setting as essential to the purpose of tracking and assessing development trajectories (Peel, 2005). Precaution enters as the balance, where our lack of knowledge concerning particular environmental effects encourages restraint when examining the human drivers that motivate change. Caution, in light of the potential for irreparable environmental damage, does not preclude the use or importance of sustainability indicators. Rather, it challenges preconceived visions of an environmentally sustainable system and recognizes the need to relax strict adherence to those indicators and how performance targets are established. Performance target will necessarily incorporate substantial margins of error and uncertainty will conspire to reduce their accuracy over time. Decision-maker confidence is likely to erode without applying caution. Precisely when to call upon precaution as a strategy or decision-hedge can be difficult to ascertain. While there is no predetermined solution, a set of conditions have been identified that offer some guidance (van de Sluijs, 2007). Four general situations have been noted that inform the environmental sustainability question:

1. Situations where there are major uncertainties
2. Situations where there is some evidence of possible harm or environmental impact
3. Situations where the potential for damage to the environmental system is significant and irreversible
4. Situations where uncertainty cannot be reduced without creating the potential for greater harm or damage

The confounding nature of uncertainty supports the need for precaution. While precaution alone cannot insure against a poor decision, it can stimulate innovative thinking and improve the management of unknowns (Crumbling et al., 1997). Precaution may also help set priorities and facilitate "what if" deliberations as trends, impacts, and agendas are carried forward into the future. In this context, environmental sustainability expands of something more than a vague destination, but rather a prediction embedded in a plan.

3.3 Prediction as Process

Environmental sustainability is viewed in a popular context as a nonspecific destination. Although in a general sense this vision is reasonable, when larger, more practical issues are considered, the policies aimed to realign social, economic, and development agendas are considered, environmental sustainability reflects a hypothesis regarding how resource use and long-term environmental functioning can be made more compatible. The sustainability hypothesis is articulated as a plan where a set of assumptions and implementation strategies are assembled into a plan designed to achieve specific societal goals. Formulating a sustainability plan and describing the elements required to address the motivating issues explain a prediction that serves two important functions (Sarewitz and Pielke, 1999). First, prediction is a test of scientific understanding. In this context, prediction occupies a position of authoritativeness where a "hypothesis" is explored by comparing an expectation with an actuality. Second, prediction serves to support decision-making by offering a glimpse of the future that might persuade anticipatory actions.

A prediction as it relates to the question of sustainability is essentially a statement pertaining to the future status of a system. That statement attempts to articulate how one believes and assumed event will transpire based on existing knowledge. Although the terms "prediction" and "forecast" are often used interchangeably, prediction suggests an expectation of a given outcome, while forecast or projection may cover a range of possibilities in a manner that is more dependent on stated assumptions. Prediction is typically depicted as a process that is considered a necessary part of complex decision-making. As a process it may vary with respect to rigor; however, the purpose remains the same: to reduce the uncertainty inherent to some future condition (Peng et al., 2013). Perhaps at its more fundamental level are those forms of informal prediction that are frequently based on informed guesses of opinion. While not as rigorous a means of defining possible future developments, informal prediction may be inductively valid if it is based on substantive knowledge, sound reasoning, and accurate data. Contrasting this "organic" style of prognostication are those science-based methods of prediction that utilize more formal and systematic means to describe future events. Relying on formalisms from the fields of statistic or mathematical modeling, scientific prediction attempts to craft a more rigorous quantitative statement concerning future events under specific conditions. One example of this style of "knowing the future" is predictive inference (Geisser, 1993).

Predictive inference emphasizes the application of past observations to guide the prediction of future observations. This approach focuses analysis on the observable parameters of an event as the foundation for deriving a logical sequence of actions that would follow from an assumed premise. The desire to embark on development pathways that conform to the ideal of environmental sustainability establishes the rationale for this type of prediction. Precisely how that prediction is easily misunderstood. The methods employed to formulate predictions vary. Perhaps the most basic sequence of activities that frame the procedure in hopes to arriving at dependent explanations of the future includes

1. Collecting data relevant to the observed system of interest
2. Searching for patterns of behavior or explicit characteristics that define system process
3. Developing statements (predictions) about future observations based on existing patterns
4. Testing the predictions (statements) and examining the outcome

Therefore, when making predictions regarding environmental sustainability, consideration of the possible patterns of both environmental factors and development together with current observation of those relationships plays a central role in the process (Figure 3.3). In the absence of theory, prediction can reduce the need for an "educated" guess, but is still something that is falsifiable. Falsifiable is important, since it means that

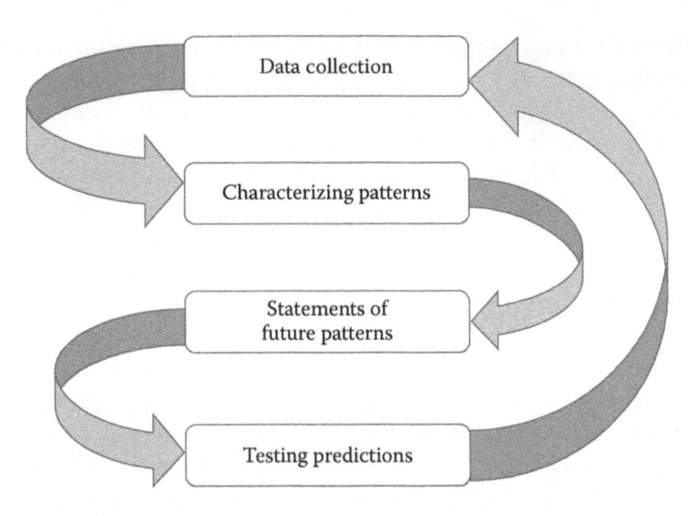

FIGURE 3.3
Stages of predicative inferences.

a prediction can be shown to be wrong, which allows for a rethinking of various strategies.

As the facet of the decision-making process and an information source to guide the implementation of development policies, predicting environmentally sustainable outcomes requires a means to examine critical decision pathways and their possible influences within the environmental system. Examination would center on the degree to which the course of a decision reduces the intensity of human disturbance with the natural environment. During the initial stages of framing a prediction, the event tree can be an extremely useful devise to trace a casual chain of events that may unfold from a decision and illuminate its potential consequences. In the absence of theory, the event tree serves as a way to test assumptions and define critical features that can be used to assist predictive modeling.

3.4 Event Tree Analysis

Predictions are enticing. They suggest confirmation of theory and understanding of uncertain situations. Above all, prediction alludes to the possibility that the future is knowable and can be controlled. However, predictions are difficult to evaluate, and questions of accuracy often remain unresolved. Simple exploratory methodologies can offer a useful solution to the confusion that surrounds uncertainty and complexity. The event (decision) tree is a widely used analysis and support tool to model a decision (Magee, 1964; March, 1994; Clemens, 2002). When applied to the predictive aspects of policy decisions assumed to promote sustainable outcomes, the decision tree identifies how choice can form a strategy as the tree branches to an acceptable conclusion. Given the ill-defined nature of environmental sustainability, the decision tree has several advantages to encourage its use:

- They are simple to understand and interpret.
- They can be used where there is little hard data.
- Insights can be generated based on subjective-expert judgment.
- Preferences can be examined.
- New information can be added into the structures of the tree.
- It views a problem as a transparent "white box" model.

An idealized event tree is illustrated in Figure 3.4. An event tree is developed in a highly iterative manner. As the decision-maker's understanding of the problem evolves, numerous changes can be made to the tree's original structure, branching points, and implied consequence pathways. In practice, the design intention is to guide and follow the direction of choice to the possible conclusions as they relate to the problem in question. However, very large and complex trees where every possible situation is considered can be counterproductive in most circumstances. For this reason, it is helpful to view a decision tree as an influence diagram that attempts to simplify the real problem into something that is manageable and facilitates the visualization of a process. Simplification is a significant aspect of the decision modeling process. By decomposing the decision into its choice and outcomes, simplification allows for a less cluttered view of the process and helps to gain insight that might be otherwise obscured by overwhelming detail and complexity. More importantly, a simple beginning to the decision problem can be enhanced through enrichment and elaboration to capture missing details as knowledge of the problem improves. Graphically, the decision tree is constructed around two main symbols: (1) a square that is used to represent a decision node and (2) a circle that is used to represent a chance node of event. The lines connecting these symbols define branches that represent the possible options or logical paths a given decision may follow. The branches emanating from a chance node explain the possible outcomes of a given course of action. Those branches in the trees, if followed, are largely determined by circumstances that lie beyond the decision-makers' control. Frequently in the design on a decision tree, those branches leading from a circle are assigned probabilities (real or subjective) to express the decision-makers' estimation of the likelihood that a particular branch in the tree will be followed. The generic structure of the decision nodes suggested in Figure 3.4 defines the controllable factors of the problem while the chance (event) nodes explain those conditions that are uncontrollable. A third type of node shown in the diagram represents the logical conclusion of a combination of decision

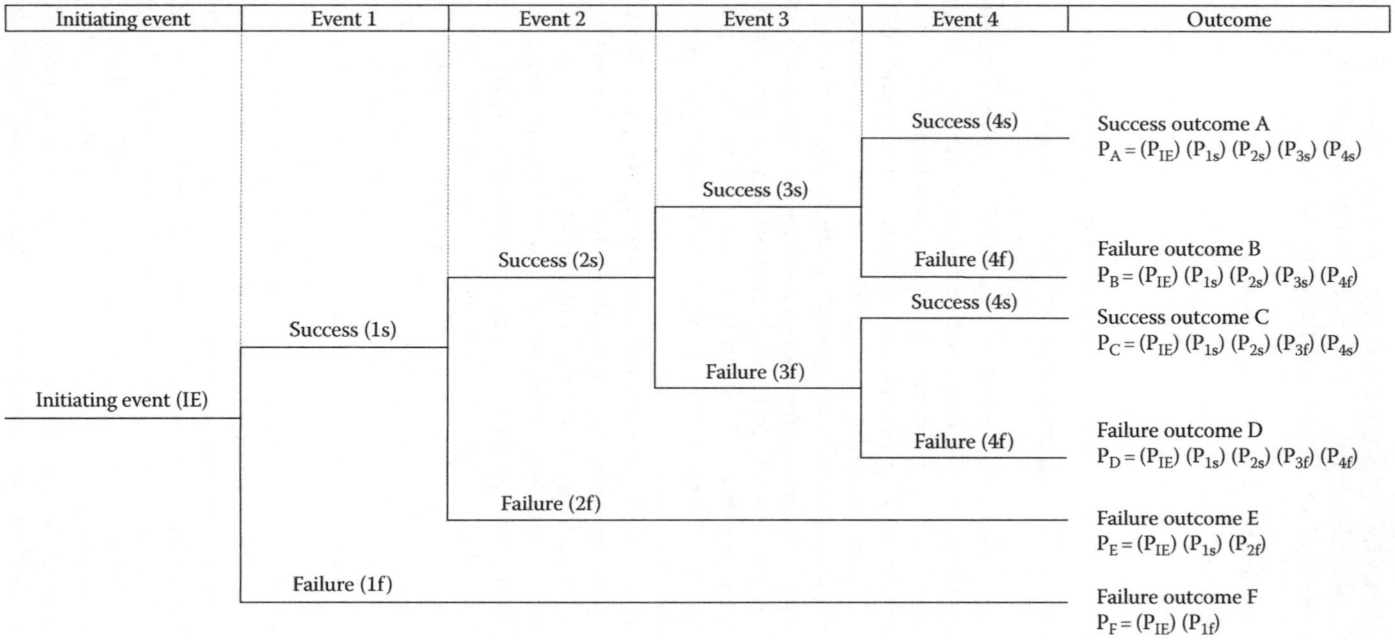

Initiating event	Event 1	Event 2	Event 3	Event 4	Outcome

Initiating event (IE)

Success (1s)

Success (2s)

Success (3s)

Success (4s)

Success outcome A
$P_A = (P_{IE}) (P_{1s}) (P_{2s}) (P_{3s}) (P_{4s})$

Failure (4f)

Failure outcome B
$P_B = (P_{IE}) (P_{1s}) (P_{2s}) (P_{3s}) (P_{4f})$

Failure (3f)

Success (4s)

Success outcome C
$P_C = (P_{IE}) (P_{1s}) (P_{2s}) (P_{3f}) (P_{4s})$

Failure (4f)

Failure outcome D
$P_D = (P_{IE}) (P_{1s}) (P_{2s}) (P_{3f}) (P_{4f})$

Failure (2f)

Failure outcome E
$P_E = (P_{IE}) (P_{1s}) (P_{2f})$

Failure (1f)

Failure outcome F
$P_F = (P_{IE}) (P_{1f})$

FIGURE 3.4
Generalized example of an event tree diagram.

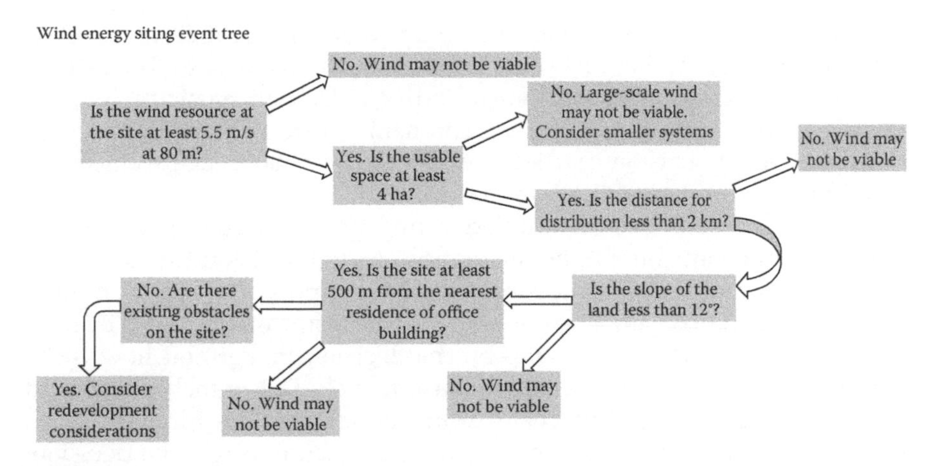

FIGURE 3.5
Illustrative example of an event tree tracing hypothetical wind energy siting decision.

events. This node is referred to as a terminal node and is often symbolized as a triangle or vertical bar to denote the end point of a decision path.

To illustrate the process of developing a decision tree relevant to the issues that surround environmental sustainability, consider the hypothetical situation where wind energy potential is being explored. In this example, the decision tree is created to guide decision-makers through the site screening process and arrive at a set of options (Figure 3.5). The tree begins with the need to determine site viability and then carries through the factors that refine choice. As the branches are followed, the options are reduced and the potential impacts are identified. At the end of each branch, the decision-maker is given the implications that may follow as that particular pathway concludes. If following a given branch is uncertain, then that decision point can be noted with a circle. Should the results require additional deliberation, then a new square is placed at that point and the options that branch from it are noted. Ideally the process continues until as many possible outcomes and decisions have been identified and explored. Once the tree displays a full logic, it can be reviewed and refined, noting solutions or possibilities that may have been overlooked.

Following the satisfactory creation of the decision tree, its contents can be evaluated. Evaluation involves assigning numeric values to each possible outcome either in the form of percentages or subjective probabilities that sum to one to capture the relative strength or influence of that branch. Although the numerical scores may be highly subjective, they offer insight into the nature of uncertainty and level of confidence that may be ascribed to a given line of reasoning. Calculating tree values to support the decision begins on the right side of the decision tree and works back toward the left. Calculation involves the simple addition of values at the nodes (decision square of uncertainty circles). Assessing the value of uncertain outcomes is accomplished by multiplying the value of the outcomes by their probability. With these values in place, the tree can be an effective way to form a conceptual view of the complex web surrounding a decision, particularly given the ill-defined aspects that define environmental sustainability. Eliciting the representations that become structured into the decision tree introduces the role of influence and causality into the solution.

Influence characterizes an implied connection between an event and the factors that contribute to its realization. As the decision tree branches, influence is the underpinning construct that supports the representation of choice and the consequence of choice. The result of choice is, in part, the product of assumptions made by the decision maker and how those assumptions may influence a given consequence. The branch points and overall paths of choice illuminate influence and can be highlighted in the tree to graphically depict how influencing factors direct the decision. Diagramming influence precedes the actual formulation of the decision tree. By focusing on influence, the dependencies that exist between events and outcomes can be noted and subjected to careful examination. Exploring influence concentrates on examining the key elements, uncertainties, and interconnections and how they are mediated over time. Through exploration, possible cause and effect relationships between variables can be defined and the larger aspects of the decision landscape brought into perspective. While decision trees and their influencing conditions are approximations of the long term and fraught with limitations that should inform their use, they offer several advantages that can support the early stages of sustainability planning by

- Providing a clear graphical depiction of a problem and its environment
- Showing potential relationships and their relevance
- Facilitating dialogue among groups with a stake in the decision
- Offering a means to compare alternatives
- Encouraging an examination of uncertainty
- Illuminating ignorance surrounding the long-term consequences associated with specific course of action

3.5 Summary

The underlying premise of this book is that environmental sustainability is a state arrived at over time. The temporal aspects of this concept together with the attempts made to project how a decision or policy may achieve some future goal highlights the presence of and frustrations created by uncertainty. Uncertainty is a characteristic of any decision and the factors of the environmental system around which sustainability agendas are organized. An understanding of uncertainty, where and how it may manifest, lends awareness to its importance and informs the environmental sustainability question by forcing a deeper consideration as to where uncertainties exist, what impact uncertain may have on a plan, what aspects of development may be irreversible, and how might the presence of uncertainty be reduced or managed. With sustainability couched as a destination, the agendas developed to achieve one or more sustainability goals serve as a prediction, and how those predictions can be understood as they are being crafted through devices such as an event tree helps illuminate the chains of potential cause and effects a decision path may invoke and thus facilitates a more explicit treatment of the unknown.

References

Abbott, J. (2005). Understanding and managing the unknown the nature of uncertainty in planning. *Journal of Planning Education and Research, 24*(3), 237–251.

Bodansky, D. (1991). Law: Scientific uncertainty and the precautionary principle. *Environment: Science and Policy for Sustainable Development, 33*(7), 4–44.

Clemens, P. L. (2002). *Event Tree Analysis*. JE Jacobs Sverdrup, Pasadena, CA.

Crumbling, D. M., Lynch, K., Howe, R., Groenjes, C., Shockley, J., Keith, L., Lesnik, B., Van, E. J., and McKenna, J. (2001). Managing uncertainty in environmental decisions. *Environmental Science and Technology, 35*(19), 404A–409A.

Deville, A. and Harding, R. (1997). *Applying the Precautionary Principle*. Federation Press, Leichhard, Australia.

Faucheux, S. and Froger, G. (1995). Decision-making under environmental uncertainty. *Ecological Economics, 15*(1), 29–42.

Geisser, S. (1993). *Predictive Inference*, Vol. 55. CRC Press, Boca Raton, FL.

Gullett, W. (1997). Environmental protection and the precautionary principle: A response to scientific uncertainty in environmental management. *Environmental and Planning Law Journal, 14*(1), 52–69.

Kriebel, D., Tickner, J., Epstein, P., Lemons, J., Levins, R., Loechler, E. L., Quinn, M., Rudel, R., Schettler, T., and Stoto, M. (2001). The precautionary principle in environmental science. *Environmental Health Perspectives, 109*(9), 871.

Lemons, J. (1997). *Scientific Uncertainty and Its Implications for Environmental Problem Solving*. Wiley-Blackwell, New York.

Lemons, J., Shrader-Frechette, K., and Cranor, C. (1997). The precautionary principle: Scientific uncertainty and type I and type II errors. *Foundations of Science, 2*(2), 207–236.

Lempert, R. J. (2003). *Shaping the Next One Hundred Years: New Methods for Quantitative, Long-term Policy Analysis*. Rand Corporation, Santa Monica, CA.

Magee, J. F. (1964). *Decision Trees for Decision Making.* Harvard Business Review, Watertown, MA.

March, J. G. (1994). *Primer on Decision Making: How Decisions Happen.* Simon & Schuster, New York.

Milliken, F. J. (1987). Three types of perceived uncertainty about the environment: State, effect, and response uncertainty. *Academy of Management Review, 12*(1), 133–143.

O'Riordan, T. (1994). *Interpreting the Precautionary Principle,* Vol. 2. Earthscan, New York.

Peel, J. (2005). *The Precautionary Principle in Practice: Environmental Decision-Making and Scientific Uncertainty.* Federation Press, Leichhardt, Australia.

Peng, G., Leslie, L. M., and Shao, Y. (Eds.). (2013). *Environmental Modelling and Prediction.* Springer Science and Business Media, New York.

Reckhow, K. H. (1994). Importance of scientific uncertainty in decision making. *Environmental Management, 18*(2), 161–166.

Sarewitz, D. and Pielke, R. (1999). Prediction in science and policy. *Technology in Society, 21*(2), 121–133.

Underwood, A. J. (1997). Environmental decision-making and the precautionary principle: What does this principle mean in environmental sampling practice? *Landscape and Urban Planning, 37*(3), 137–146.

van der Heijden, K. (1996). *Scenarios: The Art of Strategic Conversation.* Chichester, U.K.: Wiley.

Van der Sluijs, J. (2007). Uncertainty and precaution in environmental management: Insights from the UPEM conference. *Environmental Modelling and Software, 22*(5), 590–598.

Walker, W. E., Lempert, R. J., and Kwakkel, J. H. (2013). Deep uncertainty. In: Gass, S. and Fu, M. (Eds.). *Encyclopedia of Operations Research and Management Science* (pp. 395–402). Springer.

Wu, J., Jones, B., Li, H., and Loucks, O. L. (2006). Scaling and uncertainty analysis in ecology (360p). *Methods and Applications.* Arizona State University Press, Tempe, AZ.

4

Introducing the Future

In Chapter 3, the concept of environmental sustainability was described as a destination reached at some point in the future. The future is a familiar theme in environmental management and planning, particularly where there is concern for the cross-generational implications of human actions and decision on the environmental system (Allmendinger and Tewdwr-Jones, 2002). Environmental sustainability is no exception to this implicit future orientation; however, while future is a frequent keyword in policy statements regarding the environment, the integration of future thinking and the application of methods that focus attention on analytic time horizons extending beyond what the typical is not widely undertaken. Given the complexities that underscore environmental sustainability, the need for strategic foresight presents a formidable challenge for policy-makers and planners without a sound theoretical base and set of approaches to follow. In this chapter, the transdisciplinary science of futures research is introduced as a solution to the challenging aspects of environmental sustainability and the cross-generational realities embedded in this concept.

4.1 The Science of the Future

Futures research or futures studies, as it is often termed, emerged as a focus of inquiry during the late 1940s (Masini, 2006). Following World War II, interest surrounding the interrelated social, economic, and political transformations taking place during this period identified a need to explore the future consequences of the present and anticipate the trajectory of events and decision through the analysis of trends and indicators. As a focus of inquiry, futures research directs attention on the exploration of the possible long-term patterns of human development. Exploring and preparing for multiple plausible futures is a central goal of futures research. Although there is considerable debate as to whether futures research is an art or science, practitioners of this discipline seek a systematic and pattern-based understanding of the past and present in order to explain the likelihood of future events and their driving forces (Bell and Olick, 1989; Gordon, 1992; Becker, 2011).

This unique perspective is based on a series of assumptions well explained by Bell (2011) as follows:

- Time is continuous, linear, unidirectional, and irreversible.
- Not everything that will exist has existed or does exist.
- Futures thinking is essential for human action.
- In navigating our world, the most useful knowledge is knowledge of the future.
- The future is nonevidential and cannot be observed.
- There are no "facts" about the future.
- The future is not totally predetermined.
- Future outcomes can be influenced by individual and collective action.
- World interdependence invites a holistic, transdisciplinary approach to decision-making.
- Some futures models are better than others.

Conceptually, future thinking adopts an active view of decision-making (Sardar, 2010; Schuitmaker, 2011). Recognizing that decisions have long-term consequences, a futures perspective considers decisions as branch points leading away from one end and directing change toward another (Slaughter, 1993, 2002). Additionally, to the degree that it is possible to gain awareness of future consequences, means to avoid or create a specific outcome can be set into motion. The future is therefore an alternative with the implication that present choices over time direct, influence, and mobilize critical resources that enable that alternative to take shape. Future thinking places an emphasis on the role of foresight in process as a means to reduce uncertainty, anticipate change, and assume a more active role in directing change (Amara, 1991; Kristof, 2006). Foresight has become an important component of researching the future and has been used to explore the long-term direction of human action by incorporating elements of (1) futuring, which involves activities such as forecasting and forward thinking, and (2) planning to the extent that scenarios and strategic analysis combine to engage the long term (Berkhout and Hertin, 2002). Therefore, with a focus on the long term, futures research strives to gain insight not simply on possible future conditions, but also those conditions that are probable, preferable, and "wildcards" (low likelihood but high-impact events). This direction for researching the future develops a possible explanation of tomorrow through a series of intellectual and practical tasks including the following:

- *Scanning and monitoring the environment*: This process includes using both observations and research-based data to identify the direction of change, examine the relevant events acting those changes observed, and conduct an ongoing review of how change and its impact unfold.

- *Analyzing general assumptions*: Anticipatory decision-making considers external driving forces as well as the assumptions related to trends, priorities, and the resources available to see a given outcome.
- *Formulating descriptions of emerging trends*: The data-driven intelligence and the analysis of assumptions are used to create descriptions of the future from most desirable to least for each trend where change is likely to occur.
- *Compiling the forecasts*: Forecasts are based on an "if-then" process of reasoning that take past knowledge and logic to show that "if" an initial condition holds and the trend is dominant, "then" a particular outcome can be expected.

These elements of anticipatory decision-making suggest a method to examine the long term. However, critical aspects of that method should be understood along with their limitations.

4.2 Forecasting Preliminaries

Futures research, defined as the systematic exploration of what might be, does not pretend to be a predictive science (Robinson, 1990; Niiniluoto, 2001; Gordon et al., 2005). Rather, through the careful analysis of conditions and trends, alternative futures may be forecast. Therefore, researching the future is more about the task of identifying and examining the significant futures that may unfold at any time or pace. Throughout this intellectual activity, the goal is to focus on those preferred futures that can be envisions, invented, implemented, evaluated, revised, and reenvisioned. The value of the information gained from this exercise contributes to subsequent rounds of strategic planning, which in turn influences the nature and direction of operational decision-making. The careful distinction between prediction and forecasting is nontrivial (Rescher, 1998). Prediction implies certainty and that a coherent model of process exists behind the data. A prediction is assumed to be absolute and error free. Experience, however, shows that prediction is difficult if not impossible when and where uncertainty is large. For any system or relationship, there are generally more unknowns than there are soluble equations (Hare, 1985). A forecast, by contrast, is a statement or estimate of a future event that is arrived at by systematically combining the casting forward data about the past in a predetermined way. It is simply a state about the future and relies on a history of data to facilitate projection across time. A forecast also accepts uncertainty and encourages the practice of continual review and refinement, rather than the exhaustive search for perfection. When applied to the environmental sustainability question, forecasts about the future fall into two broad classes (Gordon, 1992):

1. *Exploratory*—defining forecasts of future conditions that seem plausible
2. *Normative*—explaining forecasts of future conditions that seem desirable

Regardless of class, forecast and the process of forecasting are not without limitations. Because forecasting employs data or experience of the past, it becomes less reliable the further into the future it extends. Therefore, the relative accuracy of a forecast decreases as the time horizon increases. Furthermore, both exploratory and normative forecasts can be produced using qualitative or quantitative approaches. When employed to explore sustainable outcomes

- Forecasts can be very precise, but extremely inaccurate.
- Extrapolation into the future will eventually be wrong. While extending historical trends forward is appealing, the process suggests that nothing new will deflect the trends. That contributes to the assumption that the only forces shaping tomorrow are those that exist at present.
- Forecasts by nature are incomplete.
- Plans based on forecasts must be dynamic since forecasts can be expected to be complete. Therefore, as new information is obtained, forecasts are revised and the plans based on those forecasts reviewed and amended accordingly.
- Because the future is often the product of choice, there are many possible futures.
- Forecasting is not value free.
- Forecasts may be self-fulfilling or self-defeating.

With these points noted, the process of producing a forecast can be outlined (Figure 4.1). In general, forecasting involves six fundamental tasks:

1. Determining the purpose
2. Establishing the time horizon
3. Obtaining relevant data
4. Selecting a method
5. Producing the forecast
6. Monitoring the results

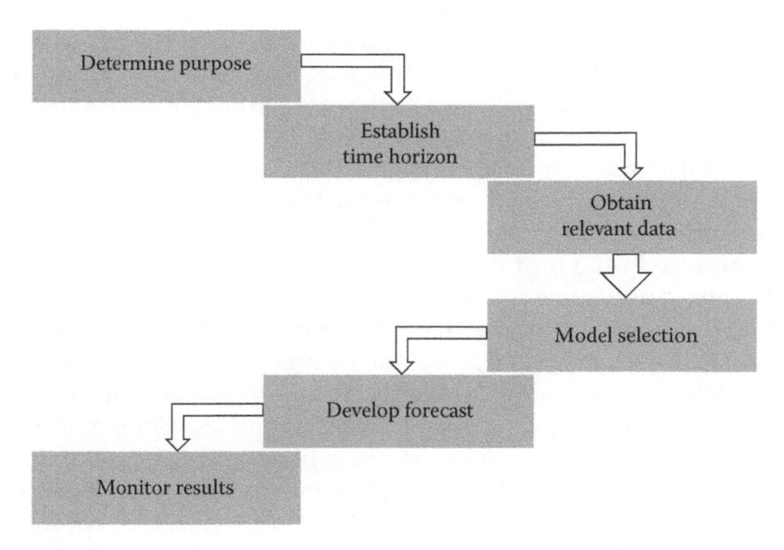

FIGURE 4.1
The general stages of the forecasting process.

Ideally, the product of this systematic approach should produce timely information expressed in a meaningful way. Because forecasting is concerned with explaining what the future will look like, the forecast serves as a valuable input into the planning process with an explicit need to examine the long term. Environmental sustainability is also a long-term forecast as defined. Deriving the long-term solution places emphasis on the capacity to forecast the direction an assumed sustainable design correlates with a sustainable environmental system (Gardner and Makridakis, 1988; Gonzalez, 1992; Swart et al., 2004). The nature of this long-term focus is explored in the section to follow.

4.3 Considering the Long Term

The relationship between futures research and planning has been widely examined (Shani, 1974; Khakee, 1988; Myers and Kitsuse, 2000). While both subject areas share an obvious future orientation, they are processes distinguished by purpose and context. Planning generally explains a systematic approach to decision-making that concentrates on the preparation of programs developed to implement specific goals and objectives. Central to the planning model is the task of measuring performance toward the attainment of those goals. Futures research strives to clarify the range of possible futures and create images (descriptions) of attainable and desirable tomorrows (Khakee and Strömberg, 1993; Schnaars, 1987; Sinclair et al., 2008;

Tonn and MacGregor, 2009). Planning may be further distinguished from futures research by its products. The plan, developed through the planning process, identifies a series of directives that are recommended for implementation by an organization, agency, or group. While futures study creates a knowledge base from which policy alternatives can be evaluated, it does not offer specific policies to arrive at some future state. Planning is, therefore, action oriented, but does not provide the broad perspective of alternative futures produced from the futures research mode (Abbott, 2005).

Time is also a noticeable contrast that separates futures research and planning. Although time is a central theme in both processes, planning is generally associated with more pragmatic and comparatively short time horizons (Connell, 2009; Dalton, 2001). The compartmentalization of time reflects the political environment within which planning operates. Here, time acts as a constraint to both the processes of crafting plans and the overall decision-making procedures that guide plan implementation. Free from those limitations, futures research can explore very long time spans, well beyond those of specific political or developmental agendas. The articulation of time has both qualitative and quantitative implications (Khakee, 1988). In a planning context, the future is examined over the short term with attention given to quantitative and linear process. The long-term perspective common to futures research adopts a more qualitative approach and considers nonlinear descriptions of processes that produce changing conditions.

Efforts to explore the type of "extreme" long-term visioning that integrate environmental sustainability into the futures research model have been discussed in the literature (May, 1982; Tonn, 2003, 2007; Samet, 2009; Pourezzat et al., 2008; Phdungsilp, 2011). The challenge to realize this futures model of sustainability centers on the need to recast time from an abstract conceptualization defined largely in qualitative terms into a more actionable construct. Considering the extreme long term within the policy arena blends a geologic time perspective into decision-making (Tonn, 2003). This novel and uncommon approach to conceptualize the nature of a decision and its ramifications has the potential to inform rational policy-making in recognizing that

- No status quo can be maintained in an environmental system for any substantial length of time
- Maintaining stasis in the bioregion for any substantial length of time is improbable
- It is highly probable the earth will undergo catastrophic events that will impact biodiversity
- Following the pattern of the past, it can be expected that extinction events will impact large flora and higher-order mammals

Planning with a considered view of the long-wave trajectories that characterize all environmental systems expands the scope of sustainability.

By adopting strategies to encompass time horizons well beyond the typical 25–50-year window, the wider scope of sustainability encourages policies to fit according to the scale of the environmental systems they are intended to function within.

A framework for conceptualizing time in the extreme has been suggested by Tonn (2003). The proposed framework consists of four main elements that are each part of the context in which decisions are made: methodology, cognition, institutions, and culture. The rational supporting inclusion of these elements and their connections into this design is based on four observations: (1) that future decision-making is complex and requires rigorous methods to support its objectives; (2) that the process, inputs, and outputs must be understandable; (3) that the methods must also be accessible to the institutions engaged in long-term planning; and (4) that long-term planning agendas must be adopted and adapted by the decision-making culture.

1. Method—implies a logic that relies on quantitative and qualitative approaches to explore the future, illuminate uncertainty, and provide evaluative criteria that can be relevant over the long term
2. Cognition—challenges how individuals conceptualize the future with the intent of bridging understanding of the long term and to reconcile uncertainty in relation to the unknown
3. Institutions—explains the capacity of government and political systems to support and adopt long-term decision models and implement long-term strategies
4. Culture—introduces the importance of value systems and how they influence decision-making, modify cultural norms to embrace the long term, and reshape how problems are identified, needs established, and alternatives evaluated

Bringing these elements together into a focused process centered on the extreme long-term calls attention to the prospects of the 1000-year plan (Tonn, 2004). The 1000-year plan introduces a time horizon well beyond the typical scope of planning activities, yet presents a time frame that (1) compliments the scale of environmental process and (2) maintains a conceptual link to the fundamentals of resource management. The 1000-year plan contains key elements that carry long-term implications and also require extended planning horizons to address. When presented with the need to resolve environmentally sustainably outcomes, the elements of a 1000-year plan display particular relevance:

- Energy—a cornerstone in sustainable development is the future of energy and energy resources
- Land use—defining the patterns of future land development, form, density, and material fabric

- Carbon management—exploring the sources and sinks of carbon, carbon footprint, and its larger relation to global cycles
- Biodiversity—protection of habitat, the maintenance of ecosystem functions, and the balanced use of the resources
- Water—a critical factor in future development, water resource, allocation and quality, and the relationship between hydrologic processes and environmental change
- Technological hazard—explaining pollution and residual management and the internment of waste generated from existing nuclear systems, toxic materials, and future high-impact activities

Crafting the 1000-year plan requires its own unique form of analysis, a future thinking that is highly prospective. Prospective analysis is the focus of the section to follow.

4.4 Prospective Analysis

The future is a multiple of many converging factors; numerous potential futures are possible and the pathways followed are not necessarily unique. Organizing thought in order to begin the exploration of a "future" and facilitating examination of the pathways that may lead to an environmentally sustainable outcome requires a shift in conventional planning approaches. A fundamental first step along this direction starts with thinking in a prospective manner. Prospective thought helps illuminate the choices of the present that informs a potential future (Godet, 1994). The prospective approach accepts the premise that a multiplicity of possible outcomes exist at any time and the actual future will develop from the interplay between the various driving variables that define a given process (Figure 4.2). How the future of a system evolves is defined as much by human action as by the influence of causalities. Prospective thinking as the initial step in forecasting employs the realization that a future is both the product of causality, an element of chance, and the image of the future imprinted upon the present (Godet, 1994). To stimulate prospective thought, key assumptions of the present can be examined and those features of the present that influence process defined. The features of interest include the identification of

- Invariants—phenomenon assumed to be constant over the time horizon
- Trends—movements affecting a phenomenon in such a way that its developments can be forecast

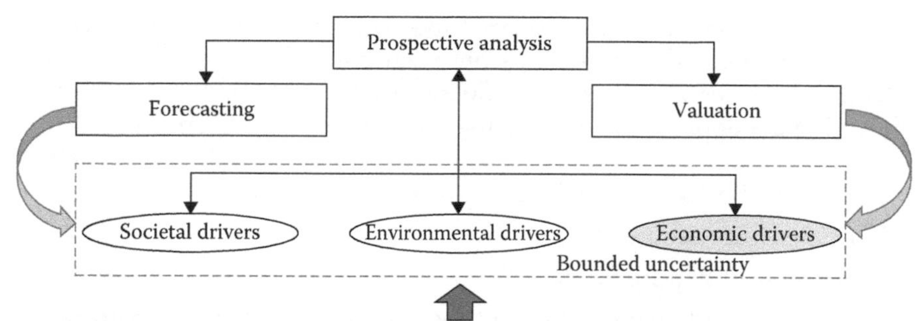

FIGURE 4.2
A typology of futures research methods.

- Germs—those nearly imperceptible factors of change that may develop as a trend in the future
- Actors—those elements that play an important role in the system through attributes of the system
- Strategies—a set of tactics or conditional decisions that explain behavior
- Conflicts—confrontations between opposing strategies
- Events—a behavior exhibited by a variable in the system
- Randomness

Careful consideration of the qualities that characterize the processes likely to populate the future suggests a direction that forecasting methods can follow. Forecasting, built on prospective analysis, explains the desire to define change. However inexact a forecast, the forecasting process is definitive, systematic, and supported by a suite of techniques that ultimately depend on human judgment and intuition. The commonly used methods of forecasting fall into two main categories: qualitative and quantitative. Qualitative techniques often provide that framework within which quantitative methods are applied. Qualitative forecasting tools attempt to assemble data, often based on subjective opinion; applying rating schemes based or subjective probabilities to that information to suggest a possible future condition. Such approaches find useful application in situations that are novel, and the amount, type, and quality of historical data are limited.

Quantitative forecasting methods require access to sufficient and appropriate base of relevant data. Typically quantitative techniques fall into methods that are statistically based or are deterministic. Statistical forecasting focuses exclusively on patterns, changes, and disturbances caused by random influences. Deterministic methods center on the identification and explicit estimation of relationships between the conditions subject to forecasting and the set of factors that influence that condition.

Selecting the appropriate forecasting technique depends on

1. The time horizon over which the forecast is to extend
2. The need for accuracy
3. The level of required detail
4. The ease with which the methods selected match the planning process

The degree to which each modifies the forecasting procedures is problem specific. Forecasting, as it relates to environmental sustainability, capitalizes on describing the range of plausible future pathways of a combined social, economic, and environmental system punctuated by uncertainty, surprise, human choice, and complexity. To assist in the forecasting and to aid in selecting the most useful method to research the future, a series of focusing questions have been offered (Myers and Kitsuse, 2000; Swart et al., 2004):

- What conceptual model best integrates the dynamic interactions between society and environment?
- Which long-term environmental and developmental trends are relevant to environmental sustainability objectives?
- To what degree can scientifically meaningful limits be defined that signal environmental degradation?
- How can the future be scanned in a manner that is creative, rigorous, policy relevant, and reflect the normative character of environmental sustainability?

The answers to these questions rely on visioning strategies.

4.5 Visioning the Future

From the previous sections of this chapter, it can be seen that discovering a future that is largely unknown and potentially unknowable is a vexing exercise. A useful means to begin thinking, forecasting, and exploring the alternative paths a future may reveal is to engage in a visioning exercise (Gaffikin and Sterrett, 2006). Visioning is the act of imagining the future and it holds promise in developing the description of a sustained human-environmental system (Olson, 1995; Shipley and Newkirk, 1998; Shipley et al., 2004; Walzer and Hamm, 2010). At its core, visioning defines a form of abstract thinking that requires decision-makers to create in their minds a "picture" of the future state they would hope to realize. Based on this image of the future, the next task is to develop the strategies that would make that future state a reality. Visioning can be used early on in both forecasting or policy setting. Visioning encourages careful consideration and deliberation regarding the objectives and motivations that direct sustainability plans. As a thought process, the "picture" or vision of the future forms a rough draft that highlights the needs identified, actions to be taken, opportunities available, and problems to resolve. Applying visioning as a means to move in the direction of an environmentally sustainable future concentrates on

- Establishing a long-term environmental vision for the region
- Identifying regional environmental assets and challenges
- Identifying the forces and trends that could impact the regional environment
- Exploring the strategies for accomplishing the environmental vision
- Identifying the benchmarks or indicators of success that express progress toward the desired future state

This vision exercise serves as a catalyst that motivates forecasting, provides a structure upon which information and ideas can be organized, a motive for deeper questioning, and a schema that advances the search for unresolved issues and concerns. With a clear vision in place, the method(s) to conduct exploratory forecasting can be selected. The methods of futures research and their strengths and limitations together with their capacities to elucidate the interpretation of alternative sustainable futures are examined in Chapter 5.

4.6 Summary

The future is an assemblage of numerous converging factors. Many potential futures are possible and the pathways followed to greet those futures are not necessarily unique. Organizing thought in order to begin the exploration of a "future" is one step to better resolve the unclear destination that is environmental sustainability. Future research offers a perspective and philosophy that facilitates an examination of the pathways that may lead to an environmentally sustainable outcome. Future thinking encourages an active view of decision-making and recognizes that decisions have long-term consequences. With a consideration of the branching points that lead from a beginning to several possible ends, futures research offers a means to apply foresight in a controlled manner to inform judgment. Through the systematic exploration of what might be, environmental sustainability becomes not unlike a long-range forecast. As plans and agendas are contemplated, visioning the extreme long term is a promising means to truly define the sustainability of a development agenda.

References

Abbott, J. (2005). Understanding and managing the unknown the nature of uncertainty in planning. *Journal of Planning Education and Research, 24*(3), 237–251.

Allmendinger, P. and Tewdwr-Jones, M. (2002). *Planning Futures: New Directions for Planning Theory*. Psychology Press, London, U.K.

Amara, R. (1991). Views on futures research methodology. *Futures, 23*(6), 645–649.

Becker, J. (2011). Evaluating a complex and uncertain future. *World Futures, 67*(1), 30–46.

Bell, W. (2011). *Foundations of Futures Studies: Human Science for a New Era: Values, Objectivity, and the Good Society*. Transaction Publishers, Piscataway, NJ.

Bell, W. and Olick, J. K. (1989). An epistemology for the futures field: Problems and possibilities of prediction. *Futures, 21*(2), 115–135.

Berkhout, F. and Hertin, J. (2002). Foresight futures scenarios. *Greener Management International, 2002*(37), 37–52.

Connell, D. J. (2009). Planning and its orientation to the future. *International Planning Studies, 14*(1), 85–98.

Dalton, L. C. (2001). Thinking about tomorrow bringing the future to the forefront of planning. *Journal of the American Planning Association, 67*(4), 397–401.

Gaffikin, F. and Sterrett, K. (2006). New visions for old cities: The role of visioning in planning. *Planning Theory and Practice, 7*(2), 159–178.

Gardner Jr., E. S. and Makridakis, S. (1988). The future of forecasting. *International Journal of Forecasting, 4*(3), 325–330.

Godet, M. (1994). *From Anticipation to Action*. UNESCAO Publishing, Paris, France.

Gonzalez, M. V. (1992). Environmental uncertainty, futures studies, and strategic planning. *Technological Forecasting and Social Change, 42*(4), 335–349.

Gordon, T. J. (1992). The methods of futures research. *Annals of the American Academy of Political and Social Science, 522*, 25–35.

Gordon, T. J., Glenn, J. C., and Jakil, A. (2005). Frontiers of futures research: What's next? *Technological Forecasting and Social Change, 72*(9), 1064–1069.

Khakee, A. (1988). Relationship between futures studies and planning. *European Journal of Operational Research, 33*(2), 200–211.

Khakee, A. and Strömberg, K. (1993). Applying futures studies and the strategic choice approach in urban planning. *The Journal of the Operational Research Society, 44*(3), 213–224.

Kristóf, T. (2006). Is it possible to make scientific forecasts in social sciences? *Futures, 38*(5), 561–574.

Masini, E. (2006). Rethinking futures studies. *Futures, 38*(10), 1158–1168.

May, G. (1982). The argument for more future-oriented planning. *Futures, 14*(4), 313–318.

Myers, D. and Kitsuse, A. (2000). Constructing the future in planning: A survey of theories and tools. *Journal of Planning Education and Research, 19*(3), 221–231.
Niiniluoto, I. (2001). Futures studies: Science or art? *Futures, 33*(5), 371–377.
Olson, R. L. (1995). Sustainability as a social vision. *Journal of Social Issues, 51*(4), 15–35.
Phdungsilp, A. (2011). Futures studies' backcasting method used for strategic sustainable city planning. *Futures, 43*(7), 707–714.
Pourezzat, A. A., Mollaee, A., and Firouzabadi, M. (2008). Building the future: Undertaking proactive strategy for national outlook. *Futures, 40*(10), 887–892.
Rescher, N. (1998). *Predicting the Future: An Introduction to the Theory of Forecasting.* SUNY Press, Albany, NY.
Robinson, J. B. (1990). Futures under glass: A recipe for people who hate to predict. *Futures, 22*(8), 820–842.
Samet, R. H. (2009). *Long-range Futures Research: An Application of Complexity Science.* 4-Scene Development Corporation Limited, Grasby, U.K.
Sardar, Z. (2010). The namesake: Futures; futures studies; futurology; futuristic; foresight—What's in a name? *Futures, 42*(3), 177–184.
Schnaars, S. P. (1987). How to develop and use scenarios. *Long Range Planning, 20*(1), 105–114.
Schuitmaker, T. J. (2012). Identifying and unravelling persistent problems. *Technological Forecasting and Social Change, 79*(6), 1021–1031.
Shani, M. (1974). Futures studies versus planning. *Omega, 2*(5), 635–649.
Shipley, R. and Newkirk, R. (1998). Visioning: Did anybody see where it came from? *Journal of Planning Literature, 12*(4), 407–416.
Shipley, R., Feick, R., Hall, B., and Earley, R. (2004). Evaluating municipal visioning. *Planning Practice and Research, 19*(2), 195–210.
Sinclair, A. J., Diduck, A., and Fitzpatrick, P. (2008). Conceptualizing learning for sustainability through environmental assessment: Critical reflections on 15 years of research. *Environmental Impact Assessment Review, 28*(7), 415–428.
Slaughter, R. A. (1993). Futures concepts. *Futures, 25*(3), 289–314.
Slaughter, R. A. (2002). *New Thinking for a New Millennium: The Knowledge Base of Futures Studies.* Routledge, New York.
Swart, R. J., Raskin, P., and Robinson, J. (2004). The problem of the future: Sustainability science and scenario analysis. *Global Environmental Change, 14*(2), 137–146.
Tonn, B. E. (2003). The future of futures decision making. *Futures, 35*(6), 673–688.
Tonn, B. E. (2004). Integrated 1000-year planning. *Futures, 36*(1), 91–108.
Tonn, B. E. (2007). Futures sustainability. *Futures, 39*(9), 1097–1116.
Tonn, B. and MacGregor, D. (2009). Individual approaches to futures thinking and decision making. *Futures, 41*(3), 117–125.
Walzer, N. and Hamm, G. F. (2010). Community visioning programs: Processes and outcomes. *Community Development, 41*(2), 152–155.

5

Methodological Approaches

The directives introduced as strategies designed to achieve an environmentally sustainable future cannot be formulated without a considered exploration of the complexities that envelop the problem. Because the very nature of what it entails to "be" sustainable is uncertain, based on a set of untested assumptions, and entirely dependent on the ability to forecast current actions forward in time, useful strategies can only be developed through the careful exercise of method. The established methods used to explore and forecast alternative futures have been reviewed extensively in the futures research literature (Fowles, 1978; Gordon, 1992; Puglisi, 2001; Glenn, 2009). In this chapter, a selection of those methods of research and forecasting the future most relevant to the task of environmental sustainability planning are examined.

5.1 Exploring Sustainable Futures

Presently, considerable effort at both the regional and national scales has been expended on programs deemed to promote environmentally sustainable outcomes (Cramer, 2013; Salomone and Saiga, 2014; Smetana et al., 2015). Many of these activities draw on common sense ideals that have been touted in the environmental literature for well over 40 years. The list is long, from reducing resource consumption, greening urban landscape, repairing damaged ecosystems, curtailing human activities known to elevate environmental risk to limiting land development to modifying economic growth agendas. Each are reasonable propositions to temper the entropic influences of human activities at present, but are they truly sustainable? A city that converts derelict industrial land to urban habitat would appear to be making a substantial contribution to the expectation of environmental sustainability as opposed to the city that rehabilitates that site for community development. Both realities evidence a political commitment and require the allocation of resources to maintain. A revised building code designed to eliminate urban sprawl and increase urban densities may also be seen as an important response to the challenge of environmental sustainability. The benefits of reduced transportation demand and the environmental impacts associated with land consumption and automobile use would suggest improved environmental conditions. However, given the definition of sustainability, is this a long-term solution? What resources will be required to support either example and how will the decisions made today set into motion unintended consequences that may not become evidence for well over 75 years?

Planning solutions targeting environmental sustainability require new ways of thinking about the future. Solutions also demand a more explicit understanding of the relationship between present actions and future outcomes (Puglisi, 2001). In the simple examples presented, sustainability principles demonstrate the need to examine alternative possibilities well beyond the immediate problem. Looking past the short term to insure that natural assets are likely to be adequately managed and preserved is the ultimate goal of the "sustainable" solution. The forces that shape the natural, social, and political dynamics that surround environmental sustainability are ultimately determined in the future. Looking beyond the immediate places importance on the methods used to evaluate programs, define how plans can be implemented, and explore their potential conflicts and consequences. The purpose of futures methodologies is to systematically explore, create, and test both possible and desirable futures to improve how decisions are made and to look comprehensively at the totality of the decision problem (Glenn, 2003). The method also includes analysis of how conditions may deviate from expectations as the result of the implementation of a policy or action and not simply their consequences. Exploring futures is not a science in the classic sense of the term. The product of research depends on the methods used and the skills of the practitioners (Glenn, 2003). The application of futures methods, however, enhances anticipatory thinking, which in turn enables more effective responses to change (Cornish, 1977; Voros, 2008). Although the future remains "unknowable," the method facilitates strategic decision-making by considering a broad range of possibilities to better inform policy agendas.

Environmental sustainability develops out of a shared, multifaceted, and compelling image of the future. Regardless of how positive that image may be, if untested by future analysis, that image may direct policy toward impossible or unreasonable goals. Forecasting studies are designed to identify factors that might impede progress to the desired goal. Further justification for engaging future analysis relates to the increasing complexity and acceleration of change (Glenn, 2003). The increasing pace of change decreases the lead time between potential events and current planning. Forecasting becomes a source of early warning by expanding the time-space for analysis. However, perhaps the most compelling reason to adopt futures methods in sustainability planning associates with a changing perception of time (Glenn, 2003). Time was generally understood as cyclical during the agricultural age when the forecasting task concerned how natural rhythms influenced planting and harvesting. By the industrial age, time was viewed in a more linear fashion with a focus on forecasting progressions and technological advances propelled society forward. In an information age, the perception of time is more expansive and forecasting shifts to the question of what is possible or desirable. Futures methodologies become a useful means to explore this expanse by identifying what we don't know but need to know in order to make intelligent decisions (Armstrong, 2001). In many respects, environmental sustainability as presently conceptualized is essentially a mental model. If our mental model of what constitutes environmental sustainability fails to incorporate an image of the future, consideration of the transition from the present state to some future condition can frustrate the formulation of effective policies and an examination of assumptions that are embedded in those directives.

5.2 A Typology of Method

The methods of futures research fall into four broad categories: qualitative, quantitative, normative, and exploratory (Figure 5.1). Qualitative models typically rely on intuition, hypothesis, and judgment in delivering forecasts. A qualitative forecast may or may not be based on detailed empirical facts; rather, methods of this type appear more speculative or hypothetical. Quantitative methods employ numerical data, mathematical calculations, and equations that capture the essential behavior of process and the interactions exhibited by variables and parameters that define the system of interest. Normative methods are based on "norms" or values that focus forecasting on the question, "What future do we want?" Normative

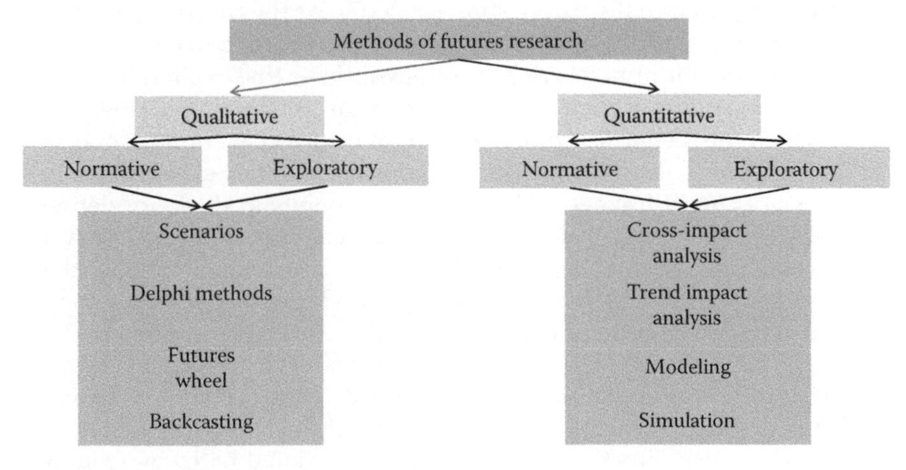

FIGURE 5.1
An example of a cross-impact matrix.

methods investigate the events that need to happen in order to realize the desired condition. These methods establish objectives, define desirable futures, and then study the possible pathways of choice and action that may lead to that future state. Exploratory methods examine what might be possible regardless of what is desirable. Forecasting generally begins from the standpoint of the present, and methods falling into this category concentrate attention on ongoing trends. Through the exploration of trends, forecasts explain what the outcome of a trend may be and how trend(s) may lead to a specific future state.

5.2.1 Qualitative Forecasting Methods

A range of techniques can be used for both normative and exploratory forecasting. The boundary that separates these methods into strictly qualitative or quantitative categories can be somewhat misleading since several techniques can fall into either class. The degree to which a technique is qualitative or quantitative depends on the availability and reliability of data together with the ability to apply statistical or mathematical models to the situation under investigation. For the purposes of this chapter, the distinction between qualitative and quantitative is used simply as an organizing framework and convenient means of reference.

- *Scenarios*: The scenario may be one of the most effective means of exploring sustainable futures (Godel and Roubelat, 1996; Armaroli et al., 2013). A scenario is essentially a story that is crafted to outline one or more conceivable/hypothetical sequence of events. As a method of futures research, scenarios are constructed primarily to focus attention on causal processes and key decision points that direct events into the future (Brewer, 2007). As the future unfolds in the scenario, patterns and groupings of processes and decision paths emerge that illuminate a rich set of possibilities that explain the possible range of outcomes that stem from an initial choice. However detailed the scenario, it functions to represent a sketch that connects a description of a specific future to present realities expressed as a series of causal linkages. The connecting points in the story define decisions and consequences that can be examined. Rather than a specific forecast, the scenario helps to organize statement about the future and brings to light the challenges and opportunities that may exit as events and trend evolve over time.

  In the context of environmental sustainability, the narrative descriptions forming a future scenario are not probability based. Their value, instead, is determined mainly by their (1) plausibility, which explains the degree to which a rational pathway from X to Y captures causal processes and explicit decisions; (2) internal consistency, which characterizes the capacity for alternative views to emerge that address similar and comparable issues; and (3) captivating content, which presents the future in a manner that is sufficiently realistic and demonstrates how that future may be affected depending on how and what decisions are made. With these three points kept in mind, the scenario is designed to systematically explore, create, and test possible and desirable future conditions. Considering the long-term commitment that envelops the concept of environmental sustainability, a scenario approach can help generate strategies and examine policy decisions extended over very long time horizons. Scenarios are also useful in many circumstances to explore and discover ignorance or the weaknesses introduced through assumptions made as the scenario unfolds.

A scenario may be exploratory or normative. Exploratory scenarios describe events and trend as they might progress based on alternative assumptions. Those assumptions are attempts to characterize how an event or trend may influence the future. Normative scenarios explain how a desired future condition may emerge from the present if certain decisions are made. Because environmental sustainability develops out of plans and policy directives, the nature of evolutionary decision path highlighted in a scenario helps to illustrate the consequence of initial assumptions and how certain directives may redirect or retard progress toward a specific sustainability goal. As a forecasting device, the scenario approach enables planners to gain an understanding of a mix of strategies, the uncertainties that surround each, and the long-term requirement that environmentally sustainable development will demand. For sustainability planning purposes, the scenario can be in a proactive setting to define those situations where trends are

- Probable but shapeable
- Probable but not amenable
- Possible and shapeable
- Possible but not amenable

Developing a scenario has been compared to the process of writing a movie script (Schwartz, 1992). In general, scenario development follows a three-phase process (Glenn, 2003):

1. *Preparation phase*: This initial step in developing a scenario focuses on defining the domain of interest or what has been termed "the scenario space." Definition of the subject area enables the analyst to list the driving forces that are important or influential in the future and may form independent directions events may follow in the scenario. Several focusing questions help refine the content and direction of the narrative. These include consideration of the nature of the study,

its scope, the time horizon involved, and the critical issues that surround decision-making. During this preparation phase, analysis concentrates on

   a. Clarifying the scenario's focal question
   b. Identifying past changes and ongoing trends
   c. Characterizing the forces driving those changes
   d. The critical uncertainties that might produce distinctly different futures
   e. "Fleshing" out the major characteristics of the narrative

2. *Development phase*: This step of the scenario design explains the key measurement variables. These measures define specific conditions that have the potential to significantly impact the outcome of the scenario. During this phase, attention is directed at developing a listing of events that shape the scenario (future), the degree to which they may impact key relationships, and influence causal process. The development phase also directs attention to forecasting. With the events and measures defined, the scenario can be "set into motion" and the narrative describing the future can be prepared.
3. *Reporting phase*: This final step in the process centers on documenting the scenario forecast. Aside from the narrative, documentation may include graphics and charts that illustrate aspects of the future explained by the scenario that presents a forecast of possible outcomes.

The challenge facing the wise use of scenarios is to craft a useful and well-focused state of the future. That statement should illuminate critical issues facing the decision-maker and guide choice along a desirable pathway. Because a comparative large number of scenarios can be created for a given situation, a clear direction at the outset and a careful discussion of the active and significant drivers reduces confusion. Providing a set of detailed long-term scenarios that describe each possible future environment together with the plausible path by which a given situation could evolve reduces uncertainty and highlights options that enhance the planning effort.

- *The Delphi method*: The Delphi technique is a method of controlled debate. The value of this method rests primarily in the ideas or images it can generate regarding possible future conditions (Linstone and Turoff, 2002). As a forecasting device, Delphi is a structured approach for soliciting and collecting expert opinion through a set of carefully focused questions. The success of the Delphi method depends on the knowledge and cooperation of the experts selected to participate in it. The technique is a useful way to estimate the likelihood and an outcome of future events, particularly in situations that cannot be well defined by quantifiable information. This method of forecasting takes its name from the Oracles of Delphi, who in ancient Greece advised people based on common sense and intuition. The application of this technique in futures research has been extensively reviewed in the literature (Rowe and Wright, 1999). Delphi is not a procedure to replace more mathematically based forecasting approaches; rather, its purpose is to use judgment where model-based methods are not possible due to the absence of data or where uncertainty requires the use of human intuition or expertise (Hsu and Sandford, 2007; Landeta, 2006).

The Delphi method is defined by four essential features that guide its application: (1) anonymity, (2) iteration, (3) controlled feedback, and (4) the statistical aggregation of group responses (Rowe and Wright, 1999). Anonymity is maintained by using questionnaires to allow experts to express opinions and judgments with social pressure. Iteration explains the cycling of the questionnaire over a number of rounds that permit experts to change their opinions and modify judgment. Iteration also fosters feedback as group responses are presented back to the experts for deliberation and more focused consideration. Summaries based on simple descriptive statistics are compiled after several rounds of questionnaire iteration and are used to produce a final group judgment or response. As a means to evaluate policy directives related to environmental sustainability, the Delphi method enables decision-makers to

- Determine or develop a range of possible alternatives
- Explore and expose underlying assumptions
- Seek out information that may generate consensus
- Correlate informed judgment across a range of disciplines and specializations
- Examine the diverse and interrelated aspects of the problem.

The role of this method in sustainability planning centers on the use and assessment of expert opinion and forecasts to build a consensus regarding the future. The Delphi process to achieve the goal of consensus building follows four main stages (Gordon et al., 2005; Gordon and Glenn, 1994; Gordon, 1994).

1. *Forming a Delphi steering team*: The steering team organizes and monitors the study. Their main purpose is to identify the problem for which a forecast is required, designing the questionnaire, and selecting the expert(s). An expert can be any individual with relevant knowledge and expertise of the problem/topic.

2. *Define the problem*: Problem definition involves a careful dissection of the issues to gain a comprehensive understanding of the critical factors and relationships involved. Because the expert needs to understand precisely what they are to comment on, definition must be clear and complete. With a well-stated problem, questions can be developed to solicit expert response. Questions range from open-ended to more specific as the process iterates. In general, questions should be kept simple and to the point.
3. *Expert selection*: Perhaps the most difficult aspect of Delphi forecasting is the selection of experts. An expert must not only have specific knowledge of the subject, but must also be willing to participate. The challenge is to identify a large number of experts such that assessment is not too narrowly based, yet not so large a number that coordination becomes difficult. The rule of thumb is to select a number that provides a good basis for forecasting given the scale and complexity of the problem and the anticipated response rate from the questionnaire rounds.
4. *Developing forecasting rounds*: The initial round of questions are poised to the experts to solicit a broad response. The questionnaires are distributed and the responses are tabulated to identify trends and patterns. The steering team uses this information to develop more detailed questions in a second-round questionnaire. Second-round questions delve deeper into the problem to clarify issues, contribute additional information, and provide a basis for subsequent rounds of questioning that will terminate once a consensus is reached.

When a consensus is reached, a view of the future exists that can be further analyzed and explored. While not an exact portrayal of the future, the Delphi responses provide insight regarding the likelihood of future events and their possible impacts.

- *Futures wheel*: The futures wheel technique is a qualitative forecasting method used to identify secondary and tertiary consequences associated with possible future trends and events (Saleh et al., 2008). The method is a form of structured brainstorming that is used to organize thinking and represents a forecasting device that offers a mechanism to think through possible impacts of current and future events and identify potential effects of a plan or strategy and show complex interrelationships. The futures wheel method begins on a blank sheet of paper and a trend, event or condition is written in the middle of the page. A series of small spokes are drawn in a wheel-like fashion from the center. Primary effects or relationships are written in circles of the first surrounding ring. Secondary consequences of each primary effect are derived from a second ring and this "rippling" of implied causality continues sequentially outward until a clear image of the implications stemming from a trend or condition emerges. Initially, the process can move ahead quickly as second-, third-, and fourth-order consequences are listed with no evaluation. As more thought is given to the overall wheel, its structure can be critiqued to produce a more realistic sequence. Alternatively, the consequence of events and trends can be explored more deliberately and the plausibility of an event, its impact, and outcome can be refined. Refinement and elaboration will yield a more reasonable sequential chain of cause and effect. Although the futures wheel method is comparatively simple and facilitates deliberation regarding complex situations, the method is only as good as the collective judgment of those participating in its creation.

- *Backcasting*: Backcasting is a technique that is often defined as the opposite to forecasting (Holmberg, 1998; Robinson, 1988). The backcasting approach essentially works from a detailed description of the desired future state backward to define the steps and strategies that would be required to realize that future. This method differs from forecasting in that forecasts attempt to explain which futures are likely to happen while the backcasting method directs attention on how to achieve a desirable future. Conceptually, backcasting is appealing particularly given the strongly normative aspects that surround the definition of sustainability (Vergragt and Quist, 2011). Backcasting discards the present, assuming that the values, services, and products of today will no longer be descriptive of the future. Therefore, instead of projecting the present forward, backcasting requires a vision of the future that is an improvement of the present. Working from that vision, the current state is assessed, gaps between the present and future are identified, and actionable plans are created to fill the gaps and move from the current state to the desired future. Ideally, the backcasting approach will produce a series of actions that follow the sustainability principles that were employed in developing the future vision. The pathways to that vision are assessed in relation to how well they agree with those design principles. Those directions that are inconsistent with the design explain futures that will not achieve the desired state.

  The backcasting method involves the identification of four critical elements:

  1. The desired end point to achieve.
  2. The obstacles, opportunities, and milestones, including the steps and setbacks from the desired end point to the present.

3. The policy requirements that draw from the milestones, obstacles, and opportunities to provide a framework for the creation of specific directives that translate into concrete actions. These actions are those designed to overcome obstacles and lead to the achievement of milestones and ultimately the desired future.
4. The strategies that define the main sequence action should follow to connect the present with that future.

   The successful application of the backcasting method will produce several useful results, including

   - A timeline detailing the actions, milestones, obstacles, and opportunities together with the desired end point and how that end point can be achieved
   - A summarizing overview in the form of specific strategies
   - A list of robust actions and alternatives available to call on

As a futures research method, backcasting is typically applied to address long-term, complex issues where: (1) there is a need to change, (2) the dominant trends contribute to the problem, and (3) the time horizon in long enough to allow considerable deliberation and experimentation.

5.2.2 Quantitative Forecasting Methods

Quantitative forecasting suggests that the processes and drivers that explain the future can be expressed using numerical data and captured based on statistical or mathematical relationships (Duin and van der Duin, 1994). The underlying assumption guiding the use of quantitative forecasting methods is that a "model" exists that describes process, and process can be summarized by a set of functionally related variables. A selection of techniques germane to environmental sustainability forecasting is presented here:

- *Cross-impact analysis*: Cross-impact analysis defines a family of techniques used to evaluate change in the likelihood and occurrence of a set of possible future events and trends (Enzer, 1972). It attempts to overcome the basic limitations of many forecasting techniques that only explain isolated events without explicit consideration of possible interactions between events and among variables. Because event sequences often display some interaction and are frequently connected in time or space, one way of capturing these interactions is to produce a mathematical description of the controlling variables that define process and how they connect. An alternative approach involves producing a detailed treatment of the probability of events and their interactions. This alternative defines the cross-impact method.

Cross-impact methods assemble a set of well-defined future events and examine the potential causal chain that the expectation or occurrence of each may have on the future (Banuls and Turoff, 2011). This technique requires the development of a model in which the causal linkages that exist among important possibilities are expressed. The model is then applied to define additional chains of possible occurrences and the degree to which the presence of each possible event alters the likelihood of future events. The goal is to distinguish a small set of important chains of event from the many possible. The separation is based on an evaluation of the probability of events in the sequence and how they may influence the long-term likelihood of other events or outcomes. Probabilities expressed in the model may be calculated from real events or based on the subjective judgment of experts. Through the assessment of these probabilities, linkages among events can be examined and presented in the form of a cross-impact matrix (Figure 5.2). In a cross-impact matrix, events that enhance change are displayed as positive numbers with the size of the number denoting the relative strength of its effect. Inhibiting impacts are assigned negative values and the no-impact case is indicated by a zero. A completed matrix lists the set of events or trends

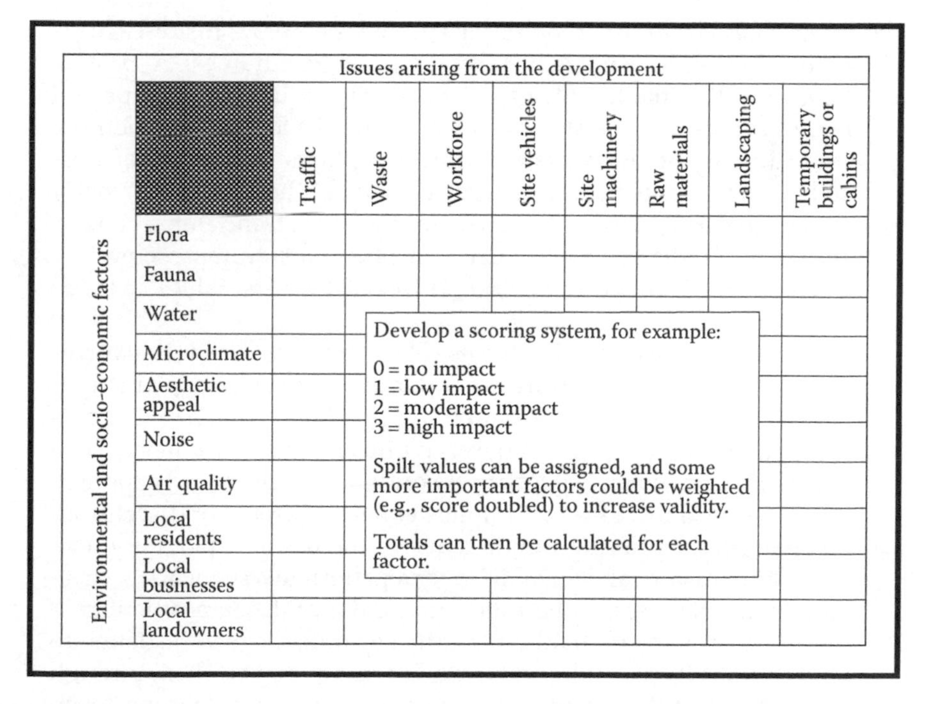

FIGURE 5.2
The general process of modeling and simulation.

that may occur across the rows while the events or trends that could possibly be affected by the row events are observed down the columns. Cumulative effects in a cross-impact chain can be evaluated according to the relations (after Enzer, 1972) shown in Figure 5.2.

Conducting a cross-impact analysis to evaluate a future condition involves seven major steps:

1. Define the events to be included in the analysis.
2. Estimate the initial probability of each event.
3. Estimate the conditional probabilities of each event pair.
4. Perform a calibration run of the cross-impact matrix.
5. Define the policies or sensitivity tests to be explored with the cross-impact matrix.
6. Perform the cross-impact calculations for the selected policies or sensitivity tests.
7. Evaluate the results and summarize the probably event pathways and relationships.

Applying cross-impact analysis within the context of decision-making has been greatly facilitated by the use of software tools (Millett, 2011).

- *Trend impact analysis*: Trend impact analysis employs historical data to produce forecasts using extrapolation techniques. A unique feature of this methodology is its capacity to include unexpected future events in an analysis. As a quantitative forecasting method, trend impact analysis systematically explores the effects of possible future events that are expected to affect the trend(s) currently subject to extrapolation. The two main goals of trend impact analysis are to (1) identify the nature of a phenomenon represented by a sequence of observations and (2) forecast future values of those observations.

  A critical issue governing the use of trend impact analysis involves the definition of a trend. In general, a trend may be explained as: (1) an inclination of an observation in a specific direction or (2) a tendency or a general direction observed in a historical sequence. Such trends may be linear, suggesting a continuous, arithmetic, exponential, or cyclical increase that implies either a seasonal or discontinuous pattern. Identifying trends requires the use of exploratory data analysis of historical data to discover a pattern in a sequence of measured variables. This often relies on the use of decomposition techniques to smooth and mathematically separate the historic data into its (1) trend, (2) seasonal, and (3) random components. Then, through the analysis of these patterns, the short-term trend defining "accidental" phenomena of limited durability and the long-term trend defining the fundamental long-term durability of the observed pattern are revealed.

Trend analysis provides useful insight regarding

- The overall pattern of change in an indicator over time
- The comparison from one time period to another
- The comparison of one geography areas to another
- The projection of the future by means of extrapolation

As a forecasting method, trend analysis (extrapolation) involves two basic procedures:

1. Using a curve-fitting algorithm to describe the historic pattern and prove a calculation of the trend into the future
2. Applying expert judgment to indentify a set of future events that, should they occur, would produce distinctive deviation in the extrapolated pattern

Trend extrapolation assumes that the phenomena under consideration are likely to continue (persist) into the future with some definable dynamic direction or rate. However, the inclusion of expert judgment modifies the assumption by allowing intervening events to impact and influence that assumed trajectory. Influence on a trend is expressed in the form of subjective probabilities that characterize the timing and expected effect of a future occurrence. High-impact future events shift the trend widely in either a positive or a negative direction from its expected pathway.

There are numerous mathematical models designed to forecast trends and cycles. Most fall under the general heading of time-trend or time series analysis and induce techniques such as (after Montgomery et al., 2015)

- Curve fitting
- Averaging methods
- Exponential smoothing
- Regression analysis
- Autocorrelation analysis and time series analysis

- *Modeling and simulation*: Modeling and simulation are techniques with a long history in environmental management and planning (Canham et al., 2003; Wainwright and Mulligan, 2005; Perry, 2009; Beven, 2010). As a forecasting method, the use of models is supported by two familiar themes: (1) the need to examine the implications of a decision in a proactive manner and (2) the desire to incorporate a number of interacting variables to form a process-view of the future. Methodologically, modeling involves two basic procedures: (1) model-building and formalization and (2) model analysis and applications. A stylized representation illustrating the basic process of modeling is presented in Figure 5.3. As suggested in this figure, modeling begins with the abstraction and simplification of an observed process of interest. That process forms a fundamental representation of the key behaviors that influence the disposition of the real-world entity the model attempts to capture. In the context of environmental sustainability, a model may describe either a replica of the system of interest or an abstraction of that system based on mathematical analogies. Constructing or applying a model, therefore, provides a systematic, explicit, and effective way to focus judgment in a structured manner. Because of the complexity associated with the concept of environmental sustainability, modeling offers a unique forecasting tool with the potential to integrate a range of factors and conditions toward the solution.

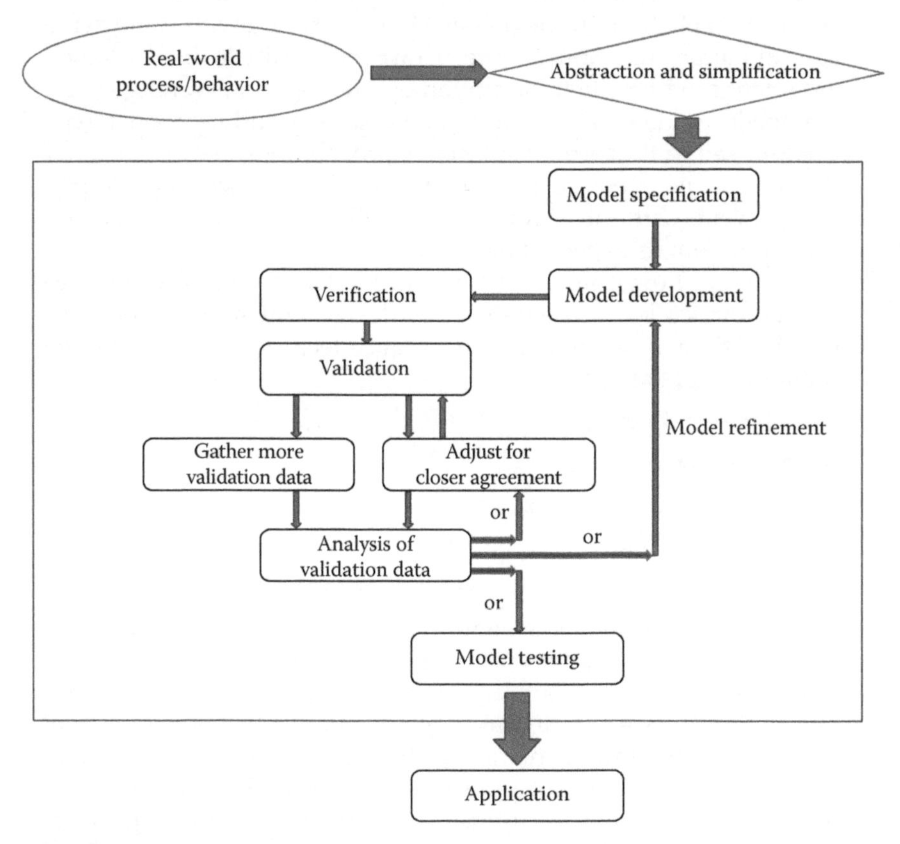

FIGURE 5.3
Geospatial analysis transforming data to information.

Developing a model of a real system and applying it in a forecasting role describes the general process of simulation (Bossel, 2013; Robinson, 2014). As an experimental and applied forecasting technique, simulation modeling is grounded in theory and observation when formulating an expression of the future characteristics of a system. Typically, simulation focuses on the development of a model and the analytical use of that model in a problem-based setting. The problem is explored through the model and examined in a controlled manner in order to reveal new insights and test prevailing assumptions. The successful application of simulation methods in environmental management or sustainability forecasting directs attention on three interrelated activities (Lein, 2003):

1. *Model design*: Model design defines the initial step in the simulation process. During this phase a detailed formulation of the problem is derived, the system under investigation is defined and the specifications to test in the model are explained.
2. *Model application*: Here the model is calibrated and the scenario or condition to be explored is selected and the mode is executed to simulate that specific case.
3. *Analysis*: Analysis is the final step in the simulation process. Analysis centers on the interpretation of the results of a model run and the overall explanation of the scenario as executed through the model.

The steps outlined suggest a basic methodology to produce information regarding possible future conditions and outcomes. That information fosters deliberation and further experimentation using the modeling approach. The specific functional form of the model and the manner by which the model represents causal processes depends largely on how the system of interest is conceptualized. There are two principal ways of capturing causality in a model: (1) continuous versus discrete processes or (2) stochastic versus deterministic. Under these broad categories, specific modeling designs have been introduced that can be adapted to the question of long-term forecasting of critical environmental, social, or economic trends that influence sustainability outcomes.

A selection of modeling schemas most relevant to sustainability forecasting includes

- *Monte Carlo sampling*: Models developed based on this strategy are used to represent systems or processes that contain stochastic elements and display behaviors that can be expressed using probabilities. Monte Carlo sampling uses artificial experience either based on data produced from a random number generator or a historic sequence of events that will form a cumulative probability distribution. That probably function is then used to simulate the behavior of future events on which that data are based.

- *Markov processes*: Markov models and their derivatives define system behavior or process characterized by the condition where the probability of a process under investigation displaying a specific status or state at a specific time is deduced from knowledge of its immediately preceding state. The Markov process explains a sequence of discrete states in time where the likelihood of the system transitioning from one state to any other is defined in relation to the calculation of each transition observed in the previous state of the system.
- *Cellular automata*: The modeling strategy can be employed to simulate how the elements of a system interact with one another. The fundamental unit of study in a cellular model is the cell. The cell explains a type of memory element that stores conditions that represent characteristics of the system under investigation. These conditions can be thought of as rules that direct and control the behavior of a cell. Based on these rules, cells exhibit distinct behaviors that propel the system to a new state or condition over time (Li and Yeh, 2000).

These modeling schemas have been employed to construct a wide assortment of software tools that can be applied in forecasting one or more aspects of the environmental sustainability question. The models listed vary in their sophistication and with respect to the data requirements needed in order to execute a simulation. Their spatial scale, scope, and complexity also vary; therefore, no single model is likely to provide a complete characterization of the future. However, models can offer useful insight into the behavior of selected critical processes that may influence future outcomes based on present conditions and existing decision strategies. For this reason alone, the model is a value tool for exploring the "what ifs" that frame decisions regarding the alternatives and uncertainties that surround the sustainability question.

5.3 Summary

Exploring the future in a systematic manner requires the careful selection of a methodology. The general methods of futures research fall into four broad categories: qualitative, quantitative, normative, and exploratory. The selection of methods that fall under each heading were introduced and the essential advantages and limitations associated with each were discussed. The methods reviewed vary in terms of sophistication and complexity with respect to application and their appropriate use will depend on the time and resources available to guide their wise use. Regardless of method, each provides a way to gain insight into the possible directions a decision or trend may follow within its environmental setting and each allows consideration of the long term with respect to a sustainability "forecast" as suggested in a plan or policy directive.

References

Armaroli, N., Balzani, V., and Serpone, N. (2013). Scenarios for the future. In: Amaroli, N. and Balzani, V. (Eds.). *Powering Planet Earth, Energy Solutions for the Future* (pp. 209–218). John Wiley & Sons, New York.

Armstrong, J. S. (2001). *Principles of Forecasting: A Handbook for Researchers and Practitioners*. Springer Science and Business Media, New York.

Bañuls, V. A. and Turoff, M. (2011). Scenario construction via Delphi and cross-impact analysis. *Technological Forecasting and Social Change, 78*(9), 1579–1602.

Beven, K. (2010). *Environmental Modelling: An Uncertain Future?* CRC Press, Boca Raton, FL.

Bossel, H. (2013). *Modeling and Simulation*. Springer-Verlag, New York.

Brewer, G. D. (2007). Inventing the future: Scenarios, imagination, mastery and control. *Sustainability Science, 2*(2), 159–177.

Canham, C. D. W., Cole, J., and Lauenroth, W. K. (2003). *Models in Ecosystem Science*. Princeton University Press, Princeton, NJ.

Cornish, E. (Ed.). (1977). *The Study of the Future: An introduction to the Art and Science of Understanding and Shaping Tomorrow's World*. Transaction Publishers, Piscataway, NJ.

Cramer, W. (2013). Regional Environmental Change refocuses on sustainability and the human-environment relationship. *Regional Environmental Change, 13*(1), 1.

Duin, P. A. and van der Duin, P. (2006). *Qualitative Futures Research for Innovation*. Eburon Uitgeverij BV, Delft, Netherlands.

Enzer, S. (1972). Cross-impact techniques in technology assessment. *Futures, 4*(1), 30–51.

Fowles, R. B. (1978). Scenarios. In: Fowles, J. (Ed.). *Handbook of Futures Research* (pp. 246–287). Greenwood Press, Hartford, CT.

Glenn, J. C. (2003). Introduction to the futures research methods series. *Futures Research Methodology, 2* (pp. 29–40). The Millennium Project, Washington, DC.

Godet, M. and Roubelat, F. (1996). Creating the future: The use and misuse of scenarios. *Long Range Planning, 29*(2), 164–171.

Gordon, T. J. (1992). The methods of futures research. *Annals of the American Academy of Political and Social Science, 522*, 25–35.

Gordon, T. J. (1994). The Delphi method. *Futures Research Methodology, 2* (pp. 1–29). The Millennium Project, Washington, DC.

Gordon, T. J. and Glenn, J. C. (1994). *Environmental Scanning*. Washington, DC: United Nations University.

Gordon, T. J., Glenn, J. C., and Jakil, A. (2005). Frontiers of futures research: What's next? *Technological Forecasting and Social Change, 72*(9), 1064–1069.

Holmberg, J. (1998). Backcasting: A natural step in operationalising sustainable development. *Greener Management International, 23*, 30–51.

Hsu, C. C. and Sandford, B. A. (2007). The Delphi technique: Making sense of consensus. *Practical Assessment, Research and Evaluation, 12*(10), 1–8.

Landeta, J. (2006). Current validity of the Delphi method in social sciences. *Technological Forecasting and Social Change, 73*(5), 467–482.

Lein, J. K. (2003). *Integrated Environmental Planning: A Landscape Synthesis.* John Wiley & Sons, New York.

Li, X. and Yeh, A. G.-O. (2000). Modelling sustainable urban development by the integration of constrained cellular automata and GIS. *International Journal of Geographical Information Science, 14*(2), 131–152.

Linstone, H. A. and Turoff, M. (2002). The Delphi Method. Techniques and Applications, Addison-Wesley, London, U.K., p. 618.

Millett, S. M. (2011). *Managing the Future: A Guide to Forecasting and Strategic Planning in the 21st Century.* Triarchy Press Limited, Devon, U.K.

Montgomery, D. C., Jennings, C. L., and Kulahci, M. (2015). *Introduction to Time Series Analysis and Forecasting.* John Wiley & Sons, New York.

Perry, G. L. (2009). Modeling and simulation. In: Castree, N., Demeritt, D., Liverman, D. and Rhoads, B. (Eds.). *A Companion to Environmental Geography* (pp. 336–357). Wiley, New York.

Puglisi, M. (2001). The study of the futures: An overview of futures studies methodologies. *Options Mediterranées, Series A, 44,* 439–463.

Robinson, J. B. (1988). Unlearning and backcasting: Rethinking some of the questions we ask about the future. *Technological Forecasting and Social Change, 33*(4), 325–338.

Robinson, S. (2014). *Simulation: The Practice of Model Development and Use.* Palgrave Macmillan, London, U.K.

Rowe, G. and Wright, G. (1999). The Delphi technique as a forecasting tool: issues and analysis. *International Journal of Forecasting, 15*(4), 353–375.

Rowe, G. and Wright, G. (2001). Expert opinions in forecasting: The role of the Delphi technique. In: Armstrong, J. S. (Ed.). *Principles of Forecasting* (pp. 125–144). Springer, New York.

Saleh, M., Agami, N., Omran, A., and El-Shishiny, H. (2008). A survey on futures studies methods. In: *Proceedings: International Conference on Informatics and Systems* (pp. 38–46). Faculty of Computers and Information-Cairo University-INFOS, Cairo, Egypt.

Salomone, R. and Saija, G. (Eds.). (2014). *Pathways to Environmental Sustainability: Methodologies and Experiences.* Springer Science and Business Media.

Schwartz, P. (1992). Composing a plot for your scenario. *Planning Review, 20*(3), 4–46.

Smetana, S., Tamásy, C., Mathys, A., and Heinz, V. (2015). Sustainability and regions: sustainability assessment in regional perspective. *Regional Science Policy and Practice, 7*(4), 163–186.

Vergragt, P. J. and Quist, J. (2011). Backcasting for sustainability: Introduction to the special issue. *Technological Forecasting and Social Change, 78*(5), 747–755.

Voros, J. (2008). Integral futures: An approach to futures inquiry. *Futures, 40*(2), 190–201.

Wainwright, J. and Mulligan, M. (Eds.). (2005). *Environmental Modelling: Finding Simplicity in Complexity.* John Wiley & Sons.

6

Geospatial Solutions

Environmental sustainability and the decisions that influence sustainable development carry an implicit spatial dimension (Backhaus et al., 2002; Peng et al., 2011). The majority of activities and programs that have been introduced to assess the broad context of sustainability, however, display a preoccupation with highly aggregated data. Data at that level of resolution often contains little information regarding the spatial patterns that underscore both environmental process and decision-making. The spatial focus introduces an activity or action-based bias that tends to overlook the obvious that "what is sustainable" must be followed by an equally detailed consideration of "where is sustainable." When attention is directed toward the spatial dimension, sustainability forecasting attempts to resolve answers to three pragmatic questions: (1) Where in the regional setting is "development" environmentally sustainable? (2) What is an environmentally sustainable development pattern? (3) How might the spatial pattern of development change under contrasting criteria and political agendas to a more sustainable disposition? Addressing these interrelated questions introduces geospatial analysis into the sustainability assessment and forecasting process.

Geospatial analysis defines a family of methods and technologies that exploit the application and analytic manipulation of geographically referenced data. The significance of the methods and technologies that rest under the geospatial umbrella is their capacity to facilitate a geographic representation of the sustainability decision problem. Through the representation of the inherently spatial aspects of the development process, the geographic characteristics of the regional setting that informs environmental sustainability can be made explicit, and the interrelationships that form between various societal objectives examined in a spatial context. In this chapter, the geospatial solution as a decision style to enhance the challenge of environmental sustainability assessment and forecasting is examined.

6.1 Technique and Technology

The acquisition, management, storage, visualization, and analysis of data that are location based define the functional elements that characterize geospatial technology. The systems that perform these tasks encompass an expanding array of automated environments that range from geographic information systems that capitalize on the analysis of spatially referenced data and mapping to the various remote sensing platforms that acquire data in the form of images created from electromagnetic energy emitted or reflected by earth surface objects. The significance of these technologies rests in their capacity to produce meaningful measurements that can be transformed into information that supports the decision-making process (Cartalis et al., 2000). The geospatial data collected and maintained by these systems feed decision-making and facilitate the spatial expression of land resources and land cover information. In addition, geospatial technology enables the collection of data at a range of geographic sales and levels of detail that can be tailored to the information needs required by a given problem. In relation to sustainability assessment and forecasting, spatial scale and detail place the decision into its proper geographic context and supplies the insight needed to address the "where is sustainable" question. Developing insight forms out of the controlled analytic manipulation and analysis of geospatial data.

Geospatial analysis explains a *multi*faceted series of operations performed on geographic data and the execution of specific methods designed to transform raw spatially referenced data into information (Brimicombe, 2009; Bonham-Carter, 2014; Burrough et al., 2015). Transforming raw data into information draws attention to the functionality designed into the geospatial instrument and its applicability to a specific analytic task (Figure 6.1). A generic listing of the functional components of a spatial data analysis system is given in Table 6.1. While the list is not an exhaustive treatment of the geospatial toolbox that can be called upon to assist spatial decision-making, it reflects the importance of the spatial domain when examining the

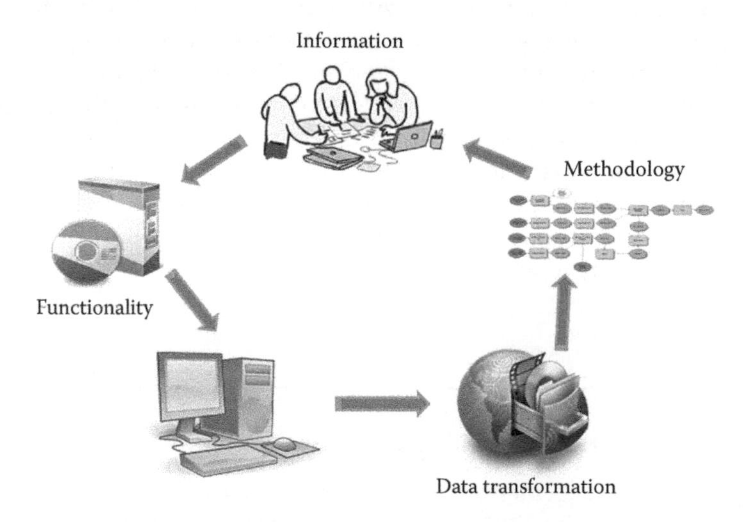

FIGURE 6.1
An example of cartographic (map) modeling workflow.

TABLE 6.1
Typical GIS Functionality

Functions	Subfunctions
Data acquisition and preprocessing	Digitizing
	Editing
	Topology building
	Projection transformation
	Format conversion
	Attribute assignment, etc.
Database management and retrieval	Data archival
	Hierarchical modeling
	Network modeling
	Relational modeling
	Attribute query
	Object-oriented database, etc.
Spatial measurement and analysis	Measurement operations
	Buffering
	Overlay operations
	Connectivity operations, etc.
Graphic output and visualization	Scale transformation
	Generalization
	Topographic map
	Statistical map
	3D bird's-eye view, etc.

long-term relationship between human development and its environmental setting. Analysis, therefore, unlocks the power of spatial thinking and guides the abstraction of geographic variables into operational patterns that can be visualized, compared, and evaluated (Lein, 2015). As demonstrated by Kropp and Lein (2012), environmental sustainability, by definition, depends heavily on the spatial representation of critical variables that explain both the human drivers that direct the development process and the biogeophysical properties that define the natural factors comprising the landscape. When examined through the lens of sustainability, a natural factor describes a physical quality of the environmental system that governs how human development will interact within the landscape (Lein, 2003). Development activities are superimposed onto this regional setting, interacting with a set of natural factors that respond to and are modified by the changes development introduces. Alternatively, a natural factor may be conceptualized as a set of physical and biological resources whose form and consequence direct the balance between land development and environmental impact.

The environmental complex is admittedly challenging to characterize. There are numerous defining attributes and interrelationships that can be identified. The capacity for geospatial methods to synthesize natural factors and illuminate their spatial distribution simplifies complexity and serves to integrate the salient features of a decision problem into a coherent representation. Geospatial simplification also allows key relationships to be tested, examined, and explored. With the interwoven actors of time, scale, and spatial variability, select characteristics of the environmental system can function as indicators. When these are placed into a geospatial framework, patterns can be demonstrated that evidence specific development scenarios (Kropp and Lein, 2012). Identifying the set of natural factors that influence sustainability potential is not well understood. A comprehensive discussion of the problem of environmental definition offered by Glasson et al. (2013) provides a reasonable starting point. The factors selected for analysis are considered "decision relevant" given the nature of the problem. These form the baseline conditions used for assessment and forecasting as the landscape is projected forward. To be relevant to the question of sustainability, the selected factors must

- Aptly characterize the environment
- Contain sufficient information to support forecasting (prescriptive analysis)
- Support the goals motivating environmentally sustainable development

Decision-relevant factors may fall into several categories and can be further refined according to the specific attributes they possess. A basic schema outlining those characteristics is given in Table 6.2. Each attribute shown is assembled into a collection that forms a representation of the regional setting. This depiction of the landscape is commonly referred to in geospatial analysis as the layer cake model (Lein, 2003). The layer cake model, serving as an abstraction of landscape qualities, is exercised using the geospatial

TABLE 6.2

Common Natural Factors Used for Environmental Characterization

Site location and topography.
Regional demography—population distribution within 10 and 50 km radius; land-use and water-use pattern.
Regional landmarks like historical and cultural heritage in the area. For this archaeological or state register can be checked.
Geology—Groundwater and surface water resources are quantified; water, quality, pollution sources, etc., are studied.
Hydrology—Groundwater and surface water resources are quantified; water, quality, pollution sources, etc., are studied.
Meteorology—Temperature extremes, wind speed and direction, dew point, atmospheric stability, rainfall, storms, etc., are recorded.
Ecology—The flora, fauna, endangered species, successional stage, etc., are enlisted.

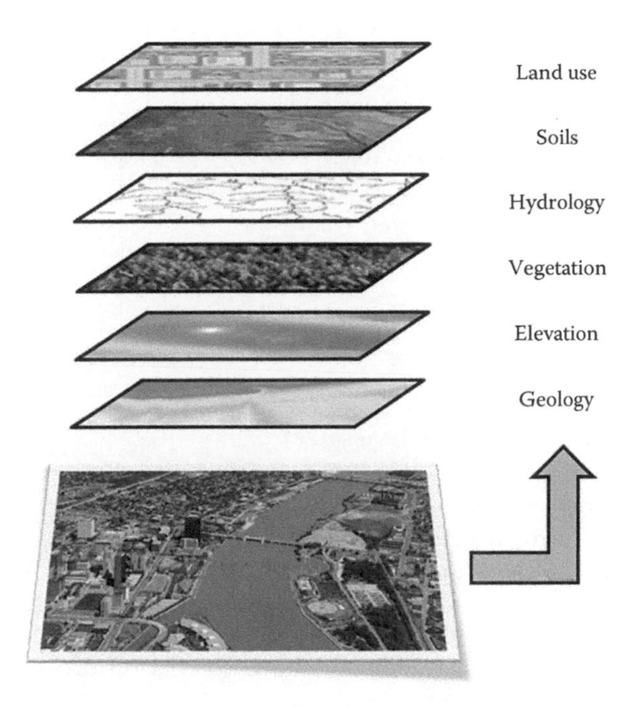

FIGURE 6.2
The general GIS layer cake model.

toolbox from which various realizations of sustainable patterns emerge (Figure 6.2). Typically, the geospatial tools employed include routines for manipulating mapped data and modeling spatial relationships. Once organized into structured methodology, spatial modeling operations can be performed according to the logic required to address a given decision problem in its geographic context.

Spatial modeling and the formulation of spatial models is a critical step in realizing the geospatial solution. A spatial model is a description of spatial phenomena that explain the basic properties of processes for a set of geographic features. Spatial models find application in the study of geographic objects and in the analysis of geographically related phenomena. Spatially explicit models useful to the analysis of environmental sustainability fall into three categories: (1) cartographic models, (2) spatiotemporal models, and (3) network models (Bolstad, 2012). Cartographic models have enjoyed a long tradition in geospatial analysis. Map realizations developed from this modeling procedure involved the application of spatial operations such as interpolation, overlay, reclassification, and conflation to mimic the logic followed when extracting and combining information from mapped data (Figure 6.3). Examples of this procedure range from the simple identification of a specific land characteristic to more involved operations that engage concepts such as habitat quality, land suitability, of locational optimization.

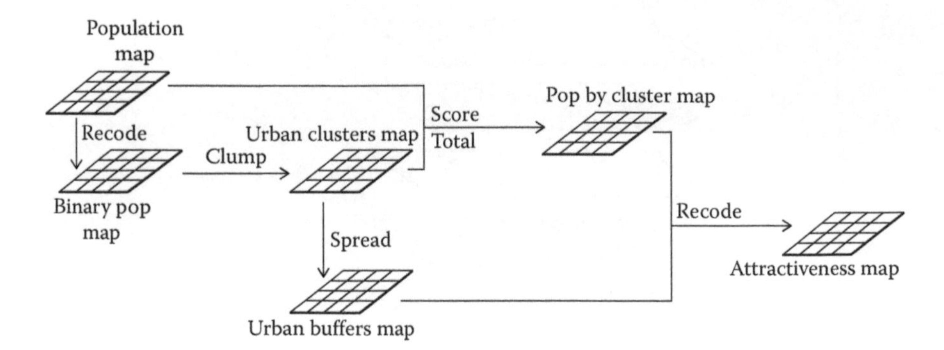

FIGURE 6.3
The logic-driven process of spatial reasoning.

Spatiotemporal models capture the dynamics of the spatial process. Models of this form incorporate time implicitly and change as a function of time-driven processes or as the result of changes altering specific spatial variables. Dispersion and flow are two common examples of spatiotemporal models. While such models are more limited in scope than the cartographic model, they offer a more mechanistic representation of processes that vary in time (i.e., land use change).

Models representing networks, such as road systems, can be designed to describe either static or dynamic processes. Network models are generally constrained to the characterization of flow or the analysis of resources defined with a set of elements organized within an interconnected system. Modeling network systems can provide useful insight regarding the costs and efficiencies related to the design of transportation networks and similar structures where there is a need to examine optimal pathways through the network, the location of infrastructure, or the expression of travel times.

6.2 Constructing Spatial Models

Environmental sustainability as a location-specific, activity-specific spatial quality returns to the underlying theme of this chapter: "where is sustainable." However, in the spatial domain, environmental sustainability is not a one-size-fits-all construct. Rather, when attempting to answer this simple question, sensitivity must be shown to the uniqueness of the regional setting and the nature, form, and function of the environmental complex that characterizes place. Assembling those qualities deemed relevant to the regional expression of development that would suggest an environmentally sustainably outcome requires the combination of a spatial data set through a sequence of operations that distill landscape features and societal needs to form either a regional expression of opportunity or constraint. Therefore, by combining multiple data layers based on sustainability principles using the geospatial toolbox, information is developed that informs decision-making (Lein, 1997).

Data combination via geospatial operations employs map algebra to guide the process. Map algebra involves the use of logical operations to translate or transform the original data into new, more useful aggregations. These new combinations of data define discrete steps in a procedure that culminate with the desired solution. The cartographic modeling procedure is an example of the map-based transformation that creates realizations of sustainability generated from the raw data considered to impart an influence on the long-term balance between human actions and environmental process. Modeling unfolds based on a set of criteria, often expressed in qualitative terms. These criteria are converted into more restrictive expressions, and as they combine they serve to filter undesirable qualities out of the decision. For example, a land cover surface may be reclassified into an expression of environmental sensitivity or a watershed system might be redefined in terms of flood potential. Both instances define a new map relationship with categories identifying areas that might restrict future development based on their connection to sustainability goals. As these layers are combined with similar intermediate expressions of the landscape, a picture forms that highlights regions of limitations or, depending on the logic, those locations where the possibility of compromising environmental integrity is minimal. Related expressions may involve requirements to be "far away from" specific landscape features or other conditions where distance zones and other geographic objects can be used to delineate surfaces that represent specific desired or necessary conditions. The fundamental logic employed in the modeling process is illustrated in Figure 6.4.

Alternatively, critical landscape variables that exhibit a dynamic quality can be examined following a spatiotemporal framework. Modeling spatial process in this manner concentrates on capturing movement or flow through the environmental system. Movement or changes in state are expressed using either deterministic or stochastic modeling strategies. Deterministic solutions focus on statistical relationships between variables to form a future explanation. Stochastic approaches introduce change through the expression of probabilistic relationships to forecast outcomes under conditions of uncertainty. In both instances, the goal is to describe the behavior of a system over time given a set of implicit assumptions. An example of this strategy can be found in Lein (2015). Recognizing that the practical analysis of environmental sustainability requires the systematic collection of temporal data, a methodology was introduced to explore the development trajectories of land use in a watershed. The land cover information generated from earth observation satellite data was coupled with a GIA-based Markov modeling procedure. Markov modeling employs a stochastic process when the probability of change from one time step to the next is dependent on the current state of the land cover pattern. This modeling approach has been successfully applied to study land use change particularly in instances where transitions in land use define active

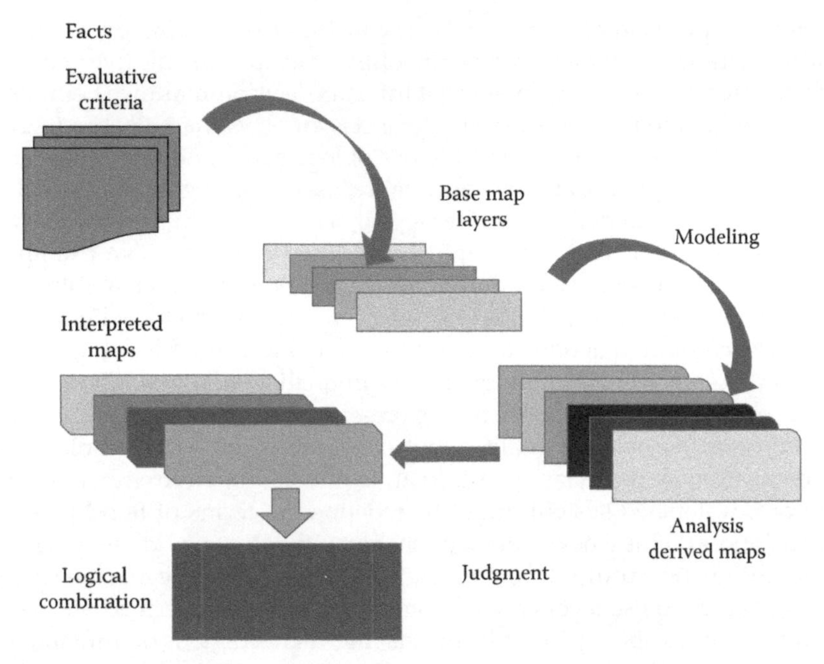

FIGURE 6.4
A typology of decision problems.

development processes that can be anticipated to alter environmental relationships. Recent examples include Du-fa (2006), Reveshty (2011), and Kumar and Mathew (2014). According to Lein (2015), Markov modeling in a GIS environment provides a simple means to examine the spatiotemporal dynamics of key indicators that reveal sustainability trends across an extended time horizon. Through this form of geospatial analysis, decision-makers can examine future spatial patterns in the land cover system in an accessible and tractable manner and use map realizations as a source of decision support. Decision support and decision analysis are two essential elements that solidify the geospatial solution. When spatial modeling is integrated into a decision support context, a robust architecture is created that lends itself to the sustainability forecasting. When connected to spatial data, the support system permits the representation and visualization of future environmental arrangements that can be compared against existing sustainability goals.

6.3 Decision Analytics

Spatial models represent a structured means of amplifying geographic patterns and producing map realizations of important decision criteria.

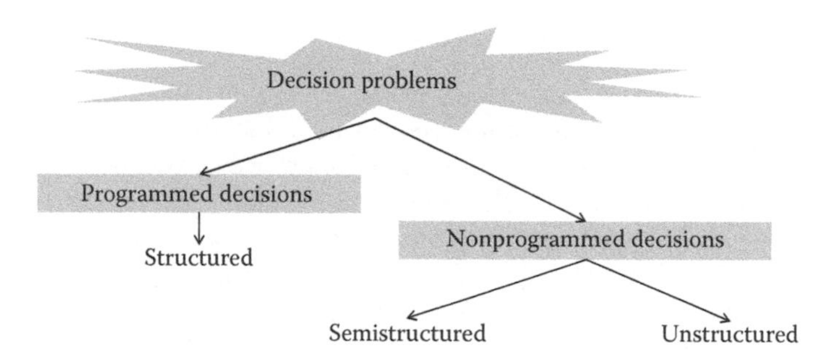

FIGURE 6.5
Three categories of decision-making situations.

The decision problem that envelops environmental sustainability and sustainability future-casting, however, is ill-defined. The strategies, indicators, and models of successful outcomes are largely unknown and reflect choice and the selection of alternatives made with a social and political context. Supporting the decision problem is a critical role for spatial modeling provided the decision problem is understood. When examined carefully, decision-making situations fall into three fundamental categories (Figure 6.5). First are the decision problems that are well defined and clearly understood. The structured problems that fall under this heading are typically addressed following a set of standard procedures and exhibit very little uncertainty since the factors and conditions involved display negligible variability. In contrary are problems that are semistructured. Here there is a developing lack of definition and an absence of standard procedures to follow. In a semistructured problem, the factors involved in the decision show greater variability and the level of uncertainty surrounding the decision is higher. Therefore, semistructured problems carry greater decision risk. Finally are those examples of decisions that are novel, highly variable, and unique. These are the unstructured decision problems that rely heavily on the intuition, judgment, knowledge, and adaptive problem-solving behavior of the decision-maker. Uncertainty permeates all aspects of an unstructured decision problem and the risks surrounding the decision are extreme.

Environmental sustainability introduces two unique problems that complicate decision-making. First, there are numerous individuals with a diversity of opinions, interests, and perspectives that make it difficult to formulate a clear set of objectives. Second, the socioenvironmental system is complex, which frustrates the reliable prediction of outcomes that may stem from a given decision or its alternatives (Reichert et al., 2015). The discovery of alternative solution and the detailed examination of outcomes under contrasting objectives and criteria can be addressed in the geospatial domain through the application of multicriteria methods of decision analysis.

6.3.1 Multicriteria Decision Analysis

Multicriteria decision analysis (MCDA) is a mechanism designed to assist the decision-maker in selecting a satisfactory alternative from a number of feasible possibilities where there is a range of choices, criteria, and priorities involved and uncertainty is high (Greene et al., 2011). This analytical methodology is a valuable means of addressing complex, ill-defined problems and has enjoyed widespread application in environmental management and decision-making (Kiker et al., 2005; Huang et al., 2011; Linkov and Moberg, 2011). In the context of environmental sustainability, MCDA facilitates the division of the decision problem (where is sustainable) into smaller and more manageable parts. By dissecting the problem into smaller subcomponents, critical features of the landscape can be studied and integrated into a meaningful solution based on preference and objectives. Multicriteria decision analysis can also be applied in group decision-making, allowing group members to discuss that decision opportunity in a manner that allows wider consideration of individual values, preferences, and concerns. With sustainability standing as an ill-defined construct, MCDA provides a way to explore "trade-offs" and rethink, adjust, and test potential solutions, particularly when faced with conflicting objectives and uncertain criteria. There are two distinctive types of MCDA problems:

- Type I—defining settings where a finite number of alternatives exist
- Type II—explaining situations where the possible solutions are infinite

Typically, environmental sustainability in the spatial domain is focused on questions associated with selection and assessment where alternatives are limited. However, even under these conditions, MCDA problems can vary widely in relation to the specifics used to frame a solution. There are common features all Type I decisions share:

- Multiple attribute and criteria that form a hierarchy
- Conflict among criteria
- A hybrid nature composed of a mix of qualitative and quantitative attributes and a mix of deterministic and probabilistic relationships
- Uncertainty stemming from a lack of complete information, data, or implicit subjectivity
- Large scale

The solution set developed from MCDA may not always be conclusive or unique. However, depending on the specifics used to define the problem, solutions fall within a range from ideal to satisfying (Greco et al., 2010; Ishizaka and Nemery, 2013). As its name implies, the ideal solution is that result that maximizes all positive criteria and minimizes those that detract from the solution. A solution is referred to as nondominated if there is no other alternative that is better than the present solution on at least one attribute and as good as it can be on any other attribute. The satisfying solution defines a reduced subset of feasible alternatives where each solution exceeds all the expected criteria. Finally are those categories of results that explain the preferred solution. Results of this type are the nondominated alternatives that best agree with the expectations of the decision-maker.

Deriving a solution set with MCDA requires selection of an appropriate approach. Although each technique is designed to make the options available to the decision-maker explicit, they differ in terms of how they combine data. Recognizing that there are many types of decisions, and the time available, amount of data to support the analysis, and the analytical skills of the decision-maker vary, the choice of method frequently depends on their

- Internal consistency
- Logical soundness
- Transparency
- Ease of use
- Data requirements

A limited selection of MCDA procedures is presented in Table 6.3. In this discussion, emphasis is given to the geographic aspects of the decision problem. This introduces a variant of MCDA referred to as spatial MCDA (Jankowski, 1995; Malczewski, 1999; Malczewski, 2006; Malczewski, 2010).

TABLE 6.3

Selected Multicriteria Decision Analysis Procedures
Analytic hierarchy process (AHP)
Analytic network process (ANP)
Best worst method (BWM)
Disaggregation—aggregation approaches (UTA, UTAII, UTADIS)
Dominance-based rough set approach (DRSA)
Evidential reasoning approach (ER)
Goal programming (GP)
Multi-attribute utility theory (MAUT)
Multi-attribute value theory (MAVT)
Stochastic multicriteria acceptability analysis (SMAA)
Value analysis (VA)
Weighted product model (WPM)
Weighted sum model (WSM)
Rembrandt method

Spatial MCDA directs attention to the problem of choice among alternatives where choice involves a set of geographically defined attributes (Sugumaran and Degroote, 2010). With attention directed to the geographic aspects of the decision, spatial MCDA involves consideration of six methodological prerequisites:

1. Definition of the goal that the decision-makers hope to realize
2. A set of preferences
3. A set of evaluative criteria that are used to examine alternative courses of action
4. Defining the set of decision alternatives
5. Examining the set of uncontrollable variables (defining contrasting states of nature)
6. Exploring the set of outcomes (possibilities) associated with each alternative

In the spatial domain, the most critical aspect of the decision problem concerns an evaluation based on the geographic distribution of attribute values and the decision-maker's preferences regarding those attribute values. The procedure used to support this evaluation is comparatively straightforward and can be conceptualized as a series of steps. In the geographic context, spatial MCDA solutions are developed following a three-phase analysis (Malczewski, 1999):

1. An intelligence phase, which takes place with a geographic information system and requires the identification, selection, and evaluation of the relevant factors that influence constraint and choice
2. A decision phase, which introduces the preferences of the decision-maker and identifies the set of feasible alternatives
3. A choice phase, where specific decision rules are applied to the data and a sensitivity analysis is performed that culminates in a recommended course of action

Therefore, a decision problem, such as the "where is sustainable" example, examined using the spatial MCDA model requires the following:

- *Establishing the decision context*: The decision context describes the administrative, political, and social structures that surround the problem. Here, the objectives are carefully stated and the criteria needed to examine the solution are defined.
- *Identifying options*: With a clear understanding of the decision environment, the set of options available are identified. These become the alternatives that form the question of choice and selection.

- *Identifying evaluative criteria*: Criteria and subcriteria are the measures of performance that will be used to judge the various alternatives. In a GIS database, these represent the set of spatial attributes (natural and human factors) that explain opportunities or constraints in relation to the decision. Evaluative criteria serve as performance measures and therefore need to be operational since these measures will specify how well the options under consideration meet the objectives. Identifying evaluative criteria is not a simple task, but they essentially become those landscape qualities that distinguish good choice from bad.
- *Establishing a scoring or weighting rubric (scheme)*: Each criterion imparts its influence on the decision. This influence is expressed in terms of a numerical scale that communicates the relative importance or degree of preference given the nature of the criterion variable and its relationship with the solution.
- *Formulating the decision rule*: The decision rule is a specific set of functions or operations that will be carried out on the data to integrate the geographic patterns of the criteria variables with the weights or preferences assigned to produce an outcome.
- *Conducting sensitivity testing*: A series of tests of the decision model is conducted to examine how the results of the analysis are affected by changes in the structure, input criteria, and weighting values employed in the decision rule. Sensitivity analysis is an exploratory procedure that helps to form a better understanding of the problem and how various elements of the problem and how various elements of the problem interact and guide the decision. Consistency of the decision and its results of analysis is the determining goal.
- *Offering recommendations*: The culmination of spatial MCDA is the recommendation. Recommendations take forms as geographic representations (maps) depicting alternatives in their spatial context. These maps illustrate the patterns future actions may produce with the landscape system. This visual product allows decision-makers to view what a given outcome may look like given the criteria and preferences that were used to construct the decision model.

When conducting spatial MCDA to resolve aspects of the environmental sustainability question, a considerable amount of trial and error can be expected. Generating new options, testing assumptions, and experimenting with criteria and their weights help to reveal alternatives and illuminate uncertainties. Realistically, there is tremendous uncertainty regarding the spatial arrangement of the future and the possible events that might produce the geography of tomorrow. Attempting to accommodate all contingencies will lead to an unmanageable situation. Therefore, a useful strategy when applying spatial MCDA is to employ a small number of bounding scenarios that can frame the range of alternatives into a more plausible set of outcomes. From this simple beginning, analysis can be repeated until a requisite model is obtained. A requisite model is one that is considered good enough to address the key issues and the possible directions that the choice may take (Phillips, 1984). Often unnecessary resources are committed to achieve only modest improvements in a model. Using the principles of parsimony as a guide, an effective decision can be obtained without extensive refinement of the model. Experience has shown that MCDA models are remarkably insensitive to minor variables in weights and are fairly tolerant of imprecision (Feizizadeh and Blaschke, 2013).

6.4 Crafting the Geospatial Solution

Several examples of MCDA applied to sustainability assessment have been introduced (Pereira and Duckstein, 1993; Mendoza and Martins, 2006; Graymore et al., 2009; Wang et al., 2009; Boggia and Cortina, 2010; Kropp and Lein, 2012, 2013; Schetke et al., 2012). In each example, a set of indicators formed criteria that were assigned weights that expressed the relative importance of each factor, or how each factor revealed a specific quality summarized as a preference. For example, Boggia and Cortina (2010) employed a set of nine environmental and nine socioeconomic indicators to rank geographic locations in relation to the technical or financial support required to support sustainable growth. Banai (2005) employed a GIS-based multicriteria evaluation to examine sustainable development on a micro-level spatial scale. The spatial criteria and priorities were based on sustainability principles that reflected the important qualities of a given locality. Factors such as density variations in population, the location of primary roads, land use mix, and modal access served as the spatial criteria that were combined with weights in a GIS to produce preference maps. These maps formed a forecasting of land sustainability.

Preference mapping was also a derivative product of spatial MCDA used to identify development options in relation to environmental sustainability in Kropp and Lein (2012) and Kropp and Lein (2013). In these examples, spatial MCDA functioned as a sustainability filter that combined macro-level factors, constraints, and weights with parcel-level attributes, constraints, and weights to suggest a range of sustainable development outcomes according to six preference scenarios. Constraining factors employed in these studies included prime farmland, floodplain avoidance, species habitat, wetlands, and open space lands while influencing factors centered on supportive criteria such as access to public mass transit, outdoor recreation, community interaction space, and other related cultural and environmental resources. The scenarios exercised via spatial MCDA represented the systematic addition or removal of factors and weights that modeled specific sustainability principles. Conditions such as emphasizing habitat protection or mass-transit access without social interaction produced a series of spatial realizations that could be evaluated against the existing land use system.

Sustainability assessment at the regional scale was also examined using spatial MCDA by Graymore et al. (2009). In this example, spatial MCDA demonstrated that a robust and sensitive methodology could be crafted to support decision-making and highlight differences in sustainability according to contrasting environmental, social, and economic conditions. Regional indicators used to explore sustainability patterns fell into three categories: (1) environmental, (2) social, and (3) economic. Spatial models were produced based on these indicators that delineated areas that were in need of initiatives to support progress toward regional sustainability goals. Further analysis of the spatial models provided insight regarding the manner by which indicators interact and the relative impact each indicator displayed on the region pattern.

The formulation and analysis of future-based scenarios is an essential ingredient of sustainability forecasting. The ultimate goal of conceptualizing, evaluating, and measuring environmental sustainability is to improve the planning process and introduce feasible solutions that can be employed to guide future development, and direct development toward more sustainable patterns. The scenario becomes the means to articulate specific environmental targets and to explore alternative ways of achieving those targets under a range of conceivable future conditions (Bryan et al., 2011). The scenario defines that plausible and integrally consistent future state. When these descriptions of the future are coupled with the geospatial methodologies outlined in this chapter, they facilitate the spatial prioritization of sustainability objectives and illuminate the geographic costs and benefits associated with a given decision, particularly under conditions of amplified uncertainty. Scenario thinking and scenario planning are explored in detail in the chapter to follow.

6.5 Summary

The "where is sustainable" is equally as important as the "when." This chapter introduced the potential for adopting geospatial solutions to the decision problem that seeks a "locational" explanation to guide environmentally sustainable development. Geospatial analysis defines a multifaceted series of operations performed on geographic data that can be organized to mimic a decision process. The geospatial solution also offers a technology that can integrate a range of attributes each with a geographic expression that can be employed to model an array of environmental conditions and arrangements.

Through the construction of these spatial models, approaches such as spatial multicriteria decision analysis can be used to explore the product of a decision and the various preferences that might be embedded in that decision process. Preference mapping allows alternative strategies to be examined, and the overall geographic "fit" of a specific development plan and its relationship with its environmental setting can be observed and evaluated. The multi-attribute and geographic dimension introduced through the geospatial solution offers a more comprehensive approach to the sustainability question by illuminating where in that landscape setting patterns of development may enjoy a long-term fit with the existing and dynamic environmental process.

References

Backhaus, R., Bock, M., and Weiers, S. (2002). The spatial dimension of landscape sustainability. *Environment, Development and Sustainability*, *4*(3), 237–251.

Banai, R. (2005). Land resource sustainability for urban development: Spatial decision support system prototype. *Environmental Management*, *36*(2), 282–296.

Boggia, A. and Cortina, C. (2010). Measuring sustainable development using a multi-criteria model: A case study. *Journal of Environmental Management*, *91*(11), 2301–2306.

Bolstad, P. (2012). *GIS Fundamentals*, 4th edn. Xanedu, Ann Arbor, MI.

Bonham-Carter, G. F. (2014). *Geographic Information Systems for Geoscientists: Modelling with GIS*, Vol. 13. Elsevier, New York.

Brimicombe, A. (2009). *GIS, Environmental Modeling and Engineering*. CRC Press, Boca Raton, FL.

Bryan, B. A., Crossman, N. D., King, D., and Meyer, W. S. (2011). Landscape futures analysis: assessing the impacts of environmental targets under alternative spatial policy options and future scenarios. *Environmental Modelling and Software*, *26*(1), 83–91.

Burrough, P. A., McDonnell, R. A., and Lloyd, C. D. (2015). *Principles of Geographical Information Systems*. Oxford University Press, Oxford, U.K.

Cartalis, C., Feidas, H., Glezakou, M., Proedrou, M., and Chrysoulakis, N. (2000). Use of earth observation in support of environmental impact assessments: Prospects and trends. *Environmental Science and Policy*, *3*(5), 287–294.

Du-fa, G. U. O. (2006). Prediction of land use and land cover patterns in recent yellow river delta using Markov chain model. *Soils*, *38*(1), 42–47.

Feizizadeh, B. and Blaschke, T. (2013). Land suitability analysis for Tabriz County, Iran: A multi-criteria evaluation approach using GIS. *Journal of Environmental Planning and Management*, *56*(1), 1–23.

Glasson, J., Therivel, R., and Chadwick, A. (2013). *Introduction to Environmental Impact Assessment*. Routledge, New York.

Graymore, M. L. M., Wallis, A. M., and Richards, A. J. (2009). An index of regional sustainability: A GIS-based multiple criteria analysis decision support system for progressing sustainability. *Ecological Complexity*, *6*(4), 453–462.

Greco, S., Ehrgott, M., and Figueira, J. R. (Eds.). (2010). *Trends in Multiple Criteria Decision Analysis*, Vol. 142. Springer Science and Business Media, New York.

Greene, R., Devillers, R., Luther, J. E., and Eddy, B. G. (2011). GIS-based multiple-criteria decision analysis. *Geography Compass*, *5*(6), 412–432.

Huang, I. B., Keisler, J., and Linkov, I. (2011). Multi-criteria decision analysis in environmental sciences: Ten years of applications and trends. *Science of the Total Environment*, *409*(19), 3578–3594.

Ishizaka, A. and Nemery, P. (2013). Multi-criteria decision analysis. *409*(19), 3578–3594.

Jankowski, P. (1995). Integrating geographical information systems and multiple criteria decision-making methods. *International Journal of Geographical Information Systems*, *9*(3), 251–273.

Kiker, G. A., Bridges, T. S., Varghese, A., Seager, T. P., and Linkov, I. (2005). Application of multicriteria decision analysis in environmental decision making. *Integrated Environmental Assessment and Management*, *1*(2), 95–108.

Kropp, W. W. and Lein, J. K. (2012). Assessing the geographic expression of urban sustainability: A scenario based approach incorporating spatial multicriteria decision analysis. *Sustainability*, *4*(9), 2348–2365.

Kropp, W. W. and Lein, J. K. (2013). Scenario analysis for urban sustainability assessment: A spatial multicriteria decision-analysis approach. *Environmental Practice*, *15*(02), 133–146.

Kumar, S., Radhakrishnan, N., and Mathew, S. (2014). Land use change modelling using a Markov model and remote sensing. *Geomatics, Natural Hazards and Risk*, *5*(2), 145–156.

Lein, J. (2015). Projecting regional sustainability trends using geospatial analytics. *Papers in Applied Geography*, *1*(2), 119–127.

Lein, J. K. (1997). *Environmental Decision Making: An Information Technology Approach*. Blackwell Science, Malden, MA.

Lein, J. K. (2003). *Integrated Environmental Planning: A Landscape Synthesis*. John Wiley & Sons, New York.

Lein, J. K. (2015). Toward a remote sensing solution for regional sustainability assessment and monitoring. *Sustainability*, *6*(4), 2067–2086.

Linkov, I. and Moberg, E. (2011). *Multi-Criteria Decision Analysis: Environmental Applications and Case Studies*. CRC Press, Boca Raton, FL.

Malczewski, J. (1999). *GIS and Multicriteria Decision Analysis*. John Wiley & Sons.

Malczewski, J. (2006). GIS-based multicriteria decision analysis: A survey of the literature. *International Journal of Geographical Information Science*, *20*(7), 703–726.

Malczewski, J. (2010). Multiple criteria decision analysis and geographic information systems. In: Greco, S., Ehrgott, M., and Figueira, J. (Eds.). *Trends in Multiple Criteria Decision Analysis* (pp. 369–395). Springer, New York.

Mendoza, G. A. and Martins, H. (2006). Multi-criteria decision analysis in natural resource management: A critical review of methods and new modelling paradigms. *Forest Ecology and Management*, *230*(1–3), 1–22.

Peng, J., Wang, Y., Wu, J., Shen, H., and Pan, Y. (2011). Research progress on evaluation frameworks of regional ecological sustainability. *Chinese Geographical Science*, *21*(4), 496–510.

Pereira, J. M. C. and Duckstein, L. (1993). A multiple criteria decision-making approach to GIS-based land suitability evaluation. *International Journal of Geographical Information Systems*, *7*(5), 407–424.

Phillips, L. D. (1984). A theory of requisite decision models. *Acta Psychologica, 56*(1–3), 29–48.

Reichert, P., Langhans, S. D., Lienert, J., and Schuwirth, N. (2015). The conceptual foundation of environmental decision support. *Journal of Environmental Management, 154*, 316–332.

Reveshty, M. A. (2011). The assessment and predicting of land use changes to urban area using multi-temporal satellite imagery and GIS: A case study on Zanjan, IRAN (1984–2011). *Journal of Geographic Information System, 3*(04), 298.

Schetke, S., Haase, D., and Kötter, T. (2012). Towards sustainable settlement growth: A new multi-criteria assessment for implementing environmental targets into strategic urban planning. *Environmental Impact Assessment Review, 32*(1), 195–210.

Sugumaran, R. and Degroote, J. (2010). *Spatial Decision Support Systems: Principles and Practices*. CRC Press, Boca Raton, FL.

Wang, J. J., Jing, Y. Y., Zhang, C. F., and Zhao, J. H. (2009). Review on multi-criteria decision analysis aid in sustainable energy decision-making. *Renewable and Sustainable Energy Reviews, 13*(9), 2263–2278.

7

Scenarios and Uncertainty

The word "scenario" has been used liberally throughout the previous chapters of this book. When considering choice, process, and uncertainty in relation to environmental sustainability, this simple concept assumes tremendous importance. The specification of the scenario designed to characterize the unknowable future establishes the conditions, connection, and assumptions through which a system may pass as process propels it forward in time. The conditions and assumptions embedded in the scenario also inform choice and define the set of possibilities that might describe the future. In this chapter, the scenario concept is dissected and the types and considerations that guide its creation are examined. From this discussion, the lingering uncertainties that punctuate the future and the manner by which uncertainty may be managed in the assessment of environmental sustainability are explored.

7.1 The Scenario Concept

A scenario, particularly in the context of future research, may be defined as an exploration of an alternative future. Built on this simple explanation is the notion of a scenario as the hypothetical sequence of events constructed for the purpose of focusing attention on causal processes and decision points (Börjeson et al., 2006; Heijden, 2011). Other definitions of the term give emphasis to the scenario as a road map or pathway of cause and effect that outline key branching points of a possible future, highlighting major determinants that influence how a future may evolve in one direction rather than another. While the literature bounds with definitions, the scenario is created for a purpose and serves as a narrative description of what might develop over time according to the ideas, assumptions, and trends that explain the present (Bunn and Salo, 1993; vanHulst, 2012). As such, the scenario is a useful mechanism to communicate and speculative thought regarding the future. The scenario provides some insight into how the future may develop, and encourages discussion and deliberation which stimulate imaginative thinking. Given these varied definitions, it can be helpful to review the essential characteristics of this construct that guides its formulation. The first distinguishing feature of the term is the recognition that scenarios are hypothetical. In essence, the scenario as narrative represents a hypothesis about the future. From this perspective, this hypothesis is something that can be explored with the understanding that the future it defines remains largely unknowable. Because there can be many futures that unfold from the present, there should be a complementing series of scenarios, each detailing different hypotheses that delineate the range of possibility. A second distinctive quality of a scenario is its generalized character. As suggested by the definitions provided earlier, a scenario is an outline. Not all of the details that might describe the future are (or can be) included. The scenario, in some respects, presents the future as a synopsis of a situation and illuminates the critical places where process leads to decisions and those decision points create pathways that the process may follow. Although generalized, the scenario, in order to be effective, captures the major determinants that have the potential to influence how the future develops. Viewed collectively, this narrative displays the possible consequence of choice and the chain of events that follow from a decision. Generalization also implies selectivity and the need to abstract a picture of causation from what is presently known. Finally, the scenario challenges its developers to be multifaceted and holistic when exploring the future. The future state of any system is the amalgam or aggregate of numerous interacting processes, contrasting and complementing forces of change, trends, and new developments that introduce opportunities as well as challenges. Identifying the salient properties of the system or problem that may be subjected to varying flows and impacts gives the scenario a dynamic quality that attempts to model the nature of the situation under consideration.

Scenarios become a device to help anticipate not simply a future, but also future needs. As the narrative develops not only does a picture of the future emerge but also the requirements or demands that the future may present. The scenario also provides direction on how to respond to those needs and encourages decision-makers to not only think beyond the obvious but to challenge convention and experiment (DeBradandere and Iny, 2001). A planning unit might construct a range of scenarios over time to make sense of diverse but interconnected environmental factors and reconcile critical uncertainties (Mietzner and Reger, 2005). Those scenarios can then be employed as a form of strategic thinking to guide the planning process. In this context, the scenario, as suggested by Ratcliffe (2002), becomes

- A presentation of alternative images rather that simply an extrapolation of existing trends
- A mix of qualitative perspectives with quantitative data
- An evaluation of sharp discontinuities
- A questioning of basic assumptions
- A creation of a learning experiences with a vocabulary for communicating complex conditions and options

A useful typology of the scenario concept that followed these bulleted points has been offered by Börjeson et al. (2006). While several typologies have been suggested in the literature, this categorization has a focus on causality and how predefined targets can be achieved: two important considerations to insure environmentally sustainable outcomes. According to this design, scenarios can be divided into three groups: (1) predictive, (2) exploratory, and (3) normative (Figure 7.1). The differences among these categories and the factors influencing their implementation are outlined in Tables 7.1 and 7.2. Predictive scenarios are created to respond and adapt to situations that are expected to occur. This category of scenario can be

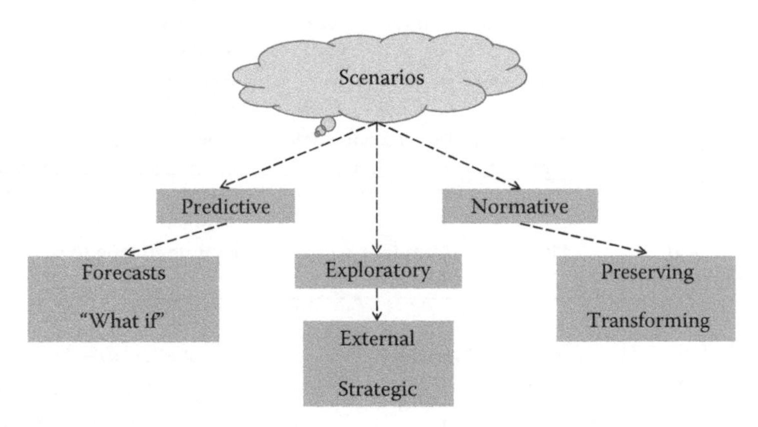

FIGURE 7.1
Elements of the scenario concept.

TABLE 7.1
Differences between Explorative and Normative Scenarios

	Explorative	Normative
Procedure	Explores possible future developments with the present as the point of departure	Identifies desirable futures or investigates how to arrive at future conditions
Function	Explorative and/or knowledge function	Target-building function and/or strategy development function
Implementation	Study of factors and unpredictabilities; test of possible actions to be taken and/or decision-making processes	Definition and concretization of goals and/or, if appropriate, identification of possible ways to reach a goal
Central question	What? What if?	How? How is it to come about? How do we get there?
Inclusion of probabilities	Possible	Indirect, part of plausible shaping and planning

TABLE 7.2

Differences between Quantitative and Qualitative Scenarios

	Quantitative	Qualitative
Implementation	When quantitative knowledge Is required And present And/or quantification is possible	When qualitative knowledge Is required Or quantitative knowledge is not present Or quantitative knowledge is not present
Topic areas	E.g., demography, economic development	E.g., institutions, culture, politics
Impact on the degree of formalization	Tendency to a high degree of formalization	Tendency to a low degree of formalization
The ideal-typical scenario technique	Modeling methods	Narrative and/or literary techniques
Manner of selecting key factors	Firm definition of a narrowly limited number of factors	Intrinsically sensory observation of details and nuances, possible without a stringent selection of factors
Chronological projection space	Short to medium term	Medium to long term

subdivided into two classes, forecasts and "what-if" scenarios. Forecasting scenarios are designed to respond to questions concerning what will happen as a situation or event unfolds. The "what-if" scenario explored the possibility that an event occurs and explains the conditions that may follow. In both instances, the goal is to gain insight regarding what is likely to happen at some point in the future. Probability and likelihood are two important components of the predictive scenario.

Exploratory scenarios are created to examine the question of what can happen and this category can be separated into external or strategic scenarios. The purpose of the explorative scenario is to examine situations or developments whose occurrence is considered possible from a variety of perspectives. To be effective, a set of explorative scenarios are developed that span the wide scope of possible developments. While conceptually similar to the "what-if" narrative, explorative scenarios differ in two important respects: (1) they are elaborative with long time horizons that permit a deep consideration of change and (2) they are often based on a starting point expressed some time in the future. With exploration as the goal, this category of scenario directs focus on how situations may develop over the long term where the pace of change may be rapid or irregular. The external scenario is one subgroup of the explorative class that concentrates attention on factors beyond the control of the decision-maker. In most cases, the external scenario attempts to inform strategy development of the planning body. The external focus provides a framework to guide policy formation and supports the development of robust strategies in direct response to external factors. The strategic scenario incorporates policy measures in its narrative. This subcategory describes the range and scope of the possible consequences of a strategic decision with a primary focus on the internal factors involved and their possible impact on the external environment. Ideally the strategic scenario describes the consequence of a decision and how a particular outcome may vary depending on which possible future unfolds. The value of the strategic narrative rests in its capacity to test different policies and their impact on target variables used to define process, which then informs strategic planning.

Reaching specific targets or goals in the future is examined through the creation of normative scenarios. Normative scenarios can fall into one of two categories: preserving or transforming. The transforming scenario addresses the problem of achieving a desired target when confronted with a prevailing (institutional or situational) structure that impedes change. Transformation in this context implies shifting to a structurally different system that induces or facilitates the changes necessary to attain the goal. A preservative scenario describes a narrative that explores how a desired target can be reached. This focus concentrates on principles borrowed from optimization, cost-efficiency, and satisficing decision-making theory to create different options that demonstrate how a specific target might be achieved. Typical, normative scenarios adopt starting points at two different stages in the decision problem, while transformative scenarios employ backcasting strategies to reverse think from a highly prioritized target, backward toward the present (Major et al., 2001; Nassauer and Corry, 2004). The backward path developed in the scenario produces a set of solutions that outline potential directions to follow that may lead to that desired targeted state. The logic is comparatively simple. If a desired future state can be clearly articulated, then the directions decision-makers will need to follow in the future in order to arrive at that state can be described in the scenario. The optimization strategies common to preserving scenarios center around the capacity to model efficient solutions. Those solutions may only be capable of capturing process over a comparatively short time horizon that constraints how deep into the future the scenario can extend.

Regardless of category, the scenario approach can suffer from a range of critical shortcomings that must be understood. These include issues related to

- A lack of diverse inputs
- A failure to gain early high-level support in their development
- An unrealistic set of goals and expectations
- A lack of specification
- The development of too many scenarios
- A failure to link the scenarios into the decision-making process
- An inappropriate time frame or scope

- A limited range of potential outcomes
- An insufficient focus on driving forces
- An overemphasis on trends
- Internal inconsistencies

Managing these potential pitfalls underscores the importance of careful scenario development and the considered selection of strategies that can be employed to construct these descriptions of the future.

7.2 Scenario Development

The scenario concept and its various instantiations suggest a methodological intent where the scenario is used to (1) delineate uncertainty space; (2) identify branching points and drives of change; (3) make explicit assumptions regarding future developments within the environment, (4) synthesize fragmented, dispersed, and vague knowledge; (5) describe a range of hypothetical directions stemming from decisions and the actions of various influences; and (6) make preparations or design strategies in response to future developments (Khakee, 1991). Several techniques for constructing scenarios have been offered and each tends to focus on four critical activities (Mietzner and Reger, 2005):

1. Defining the scope of the scenario
2. Constructing a support database
3. Crafting the scenario narrative
4. Selecting the strategic option

The specific methods introduced to create a scenario can range from simplistic to complex, qualitative to quantitative. A tractable approach to crafting a scenario that fits the demands of environmental sustainability forecasting has been suggested by De Brabandere and Iny (2010). Their method consists of nine development phases, which are summarized here:

- *Phase 1—Preparation*: This initial step concentrates on preparing the background against which the scenario will be set. Preparation requires forming a succinct definition of the problem and selection of a time horizon. This phase also involves undertaking a broad overview of the relevant trends and forming a preliminary set of hypotheses and influences that characterize the driving factors of change.

- *Phase 2—Variable selection*: By reviewing the trends, hypotheses, and influences, a set of variables can be identified that define the actors in the scenario. In the sustainability example, those variables could include factors that drive urbanization, emerging environmental threats, land resource qualities, or other processes that force the environmental system.
- *Phase 3—Conceptualization*: Conceptualization encourages creative thinking and exploration of the problem, variables, and forces diving the process acting within the environmental setting. Here, the preliminary hypotheses are examined, refined, and reimagined as the story of the future begins to take shape.
- *Phase 4—Brainstorming*: This stage in scenario development focuses on creating a framework when conceptual ideas are developed into mental models that capture the salient processes and produce contrasting hypotheses of each driving variable.
- *Phase 5—Referencing and organization*: Here, the hypotheses can be assembled into a coherent description based on their relationships, critical uncertainties, and guiding logics. A matrix is often used at this step to organize and rank the factors and trend in terms of their relative importance.
- *Phase 6—Scenario naming*: Each description of the future requires an identifying name. This name relates to the hypothesis it explains. Naming also sets the stage for developing an outline of those plausible futures that will be examined. The outline should comprise a set of three to four narratives expressing structurally or quantitatively different alternative futures.
- *Phase 7—Developing the story*: In story development, creative thought is used to enrich and augment the scenario. Enrichment adds detail and introduces consideration of the implications, challenges, and opportunities that render the scenario more convincing.
- *Phase 8—Scenario communication*: The communication phase presents the storyline portraying the future from the perspective of each scenario. These narratives, to be effective, have a structure and are presented to help the audience understand and accept each possible explanation of the future. Effective presentation introduces a sense of realism in the storyline and may include placing the scenario in a historical context, highlighting the main themes and employing illustrating depictions.
- *Phase 9—Application*: The scenario is not an action plan; it explains a set of three to four contrasting interpretations of the future. Application of the scenario focuses attention on how well the storyline promotes thought and how well it serves as a frame of reference around which policy, management, and development decisions can take shape.

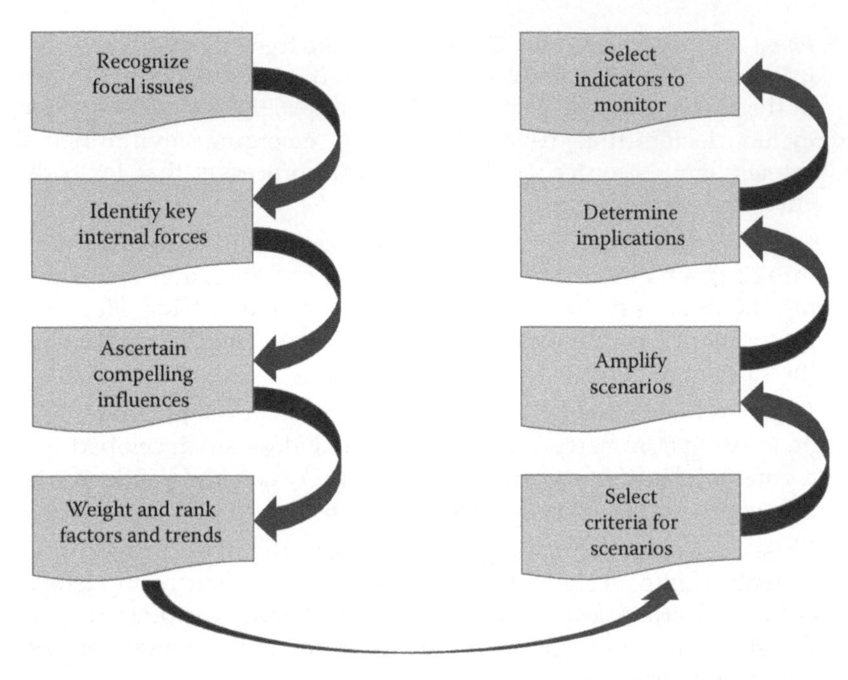

FIGURE 7.2
Eight-stage approach to scenario construction.

A similar approach to scenario construction was introduced by Schwartz (1992). This process is outlined in Figure 7.2, which suggests an eight-step sequence to scenario design. The first step centers on recognizing the focal issues that contribute to the careful identification of the forces that direct the local environment. From this stage, brainstorming is used to explore the driving and compelling influences, which then enable a ranking and subjective weighing of the key forces and drivers to express both their importance and uncertainty. From here, scenario development moves on to the selection of criteria and the formulation of the narrative. The scenario logic implicit in the narrative can be enriched through elaboration with consideration given to the modification of rankings and weights. The final two steps in the process involve (1) the determination of each scenario's implications followed by (2) the selection of indicators that will be employed to monitor the unfolding future.

In either of the two approaches presented, the central goal of scenario building is to encourage thinking in terms of time (the long term) and breadth (across a range of possible futures). Scenario generating methods are essentially techniques for producing and collecting ideas, knowledge, and perspectives regarding the possible future state of a system. The steps suggest a sequence of identifiable tasks that are performed to facilitate ideation, integration of elements, and formulation of a storyline into a coherent whole in a clear and consistent manner. The scenario represents a means to an end, and from a learning point of view, it tests our understanding as to how a future may emerge, presenting a possibility space that can be incorporated into decision-making. Only a limited number of scenarios can be developed in detail without causing the process to dissipate (Mietzner and Reger, 2005). Producing a manageable number of alternative futures in a logical manner that captures the dynamics of process can be challenging. In general, scenarios are selected for evaluation based on the following:

1. *Plausibility*: The scenario has to be capable of actually happening.
2. *Differentiation*: Each scenario should be structurally different, not simply a variation of the same theme.
3. *Consistency*: Scenarios should be logical, credible, and grounded.
4. *Decision-making value*: Each scenario should contribute useful insights into the future and be relevant to the problem.
5. *Challenge*: The future narrative should challenge conventional view regarding the future.

An extensive listing of criteria for selecting scenarios expanding on these points is given in Table 7.3. By carefully considering each item, "best practices" can be defined that may improve the quality and decision value of a scenario set. Sustainability decisions-making, with a focus on how, when,

TABLE 7.3

Criteria for Scenario Evaluation and Selection

Criteria for Scenario Evaluation and Selection
Clarity
Clear and understandable
Cohesion
Suitability
Transparency
Thoroughness of content
Flawlessness
Plausibility
Completeness
Descriptive
Precise
Relevance
Function
Useful
Identifiable
Comprehensive
Compatibility
Unique
Stable

and where to act, is currently based on present-day expectations projected into the future. The environment enveloping the sustainability question is highly unpredictable and decision-makers typically work from a limited range of expectations that run the risk of being incorrect over the long term. Therefore, even in a scenario, sensitivity to uncertainty and how it may influence the process are paramount.

7.3 Uncertainty Revisited

Uncertainty pervades all our attempts to understand the natural and physical environment (Regan et al., 2002). The sources and consequences of uncertainty are complex and often display substantial variation. When developing a scenario, the presence of uncertainty should be acknowledged but it should not preclude working through a future sustainability narrative. However, when conceptualizing and articulating those futures, categorizing uncertainty improves scenario quality. There have been several attempts to place uncertainty into perspective as it relates to the environmental process (Lemons, 1996; Abbott, 2005; Polasky et al., 2011; Zapata and Kaza, 2015). Perhaps at its most fundamental level, uncertainty branches down two parallel paths. One pathway explains the uncertainties associated with a lack of knowledge regarding the state of the physical/environmental system. Uncertainties that fall into this area have been referred to as epistemic and they develop due to limitations that stem from measurement insufficiencies of data, difficulties arising from extrapolations and interpretations, and variations that result over space and time. The second main cluster of uncertainties are termed "linguistic" that introduce themselves as a function of language (Regan et al., 2002). Natural language is populated with an array of underspecified, vague, ambiguous, and context- and culturally dependent concepts. The indeterminacies and definitional "fuzziness" of language give rise to unique forms of uncertainties that may frustrate interpretations of the future as well as the clarity by which processes and behaviors can be articulated. Both explanations of uncertainty impact scenario development, but do so in contrasting ways.

Epistemic uncertainty identifies six interrelated conditions that conspire to introduce error into the scenario outline. Recognizing their presence informs the manner by which causality is represented in the scenario and the level of detail that can be realistically employed to define process. These six conditions arise differently and impact scenario design and content:

1. *Measurement error*: Error manifests as unrecognized or uncontrolled variation in the measurement of a given quantity. This form of error may be attributed to the instruments used to measure an object or

phenomenon, or the result of the observations or sampling methods used to obtain a measure. In either instance, the measured value of the object becomes the product of its true value plus error.

2. *Systematic error*: This source develops from bias in measurement. However, it is not a random error. Systematic error can result from the judgment of analysts or failures in following sound data collection methods.
3. *Natural variation*: Dynamic environmental processes display spatial and temporal modulations that are difficult to summarize and predict. This natural stochasticity introduces a unique form of randomness that produces sources of uncertainty that may be impossible to adequately account for over the full range of spatiotemporal scales that process may define.
4. *Inherent randomness*: When considering the behavior of a system, the driving processes and patterns may be irreducible to a deterministic explanation. This condition contributes to unpredictability since the processes at work are difficult to specify in precise terms.
5. *Model uncertainty*: The manner by which a problem or system is conceptualized leads to uncertainty as conceptual representations are expressed in the form of a model. Model uncertainty introduces itself in two fundamental ways: (a) through the simplification of process and the selection of defining variables, states, and parameters, and (b) through the constructs used to express observed processes.
6. *Subjective/technical judgment*: A considerable amount of speculation and judgment surrounds future-casting in relation to environmental sustainability. Due to the absence of empirical data, an inability to quantify relationships or probabilities, or due to the lack of scientific evidence to support an interpretation, expert judgment is often called upon to fill in the gaps. While judgment-based reasoning is not uncommon, its applicability depends on the experience and observational skills of the expert(s). Variations in expertise contribute to substantial uncertainties, particularly in cases where several experts offer conflicting judgments.

Linguistic uncertainty is equally problematic in the development or application of the scenario approach. Language, definition, and interpretation issues contribute to five distinct categories of uncertainty:

1. *Vagueness*: Natural language is replete with concepts and constructs that are inexact, imprecise, and difficult to define in absolute terms. Language introduces uncertainty when meaning is inherently "fuzzy" or when definition lacks precision. Both instances lead to contrasting and often conflicting interpretations.

2. *Context*: Language introduces specific meanings that are frequently context dependent. Uncertainty arises when the context in which specific language is used is not well explained to adequately clarify a proposition. Without establishing a context, misunderstanding is possible.
3. *Ambiguity*: Because words in a language may have multiple meanings, uncertainty can develop when meaning is not well communicated.
4. *Generality*: Language concepts often lack specificity and can overgeneralize both context and meaning. Underspecifying statements can produce incomplete or overly broad interpretations leaving an element of uncertainty that can be difficult to resolve.

Uncertainty, as can be seen, is ever present and it is unlikely that it can be eliminated in the application of sustainability scenarios. However, by recognizing the common sources, it may be possible to narrow down the range of uncertainties and reduce their influence or impact on decision-making. Uncertainty may emanate from different sources, be compound, and exhibit temporal and spatial discontinuities. Even with the outline presented, uncertainty may be difficult to categorize. Nevertheless, a comprehensive and systematic treatment of uncertainty when exploring the strategies that embrace environmental sustainability principles is essential.

7.4 Scenario Planning and Visioning

Scenario development fits into the larger programs of planning and strategic forecasting. Taken in this context, scenario planning is essentially a procedure used to explore uncertainty while planning, as it relates to the question of environmental sustainability and concentrates on the allocation of functions to their appropriate spatial location. "Appropriate" implies that the placement of functions in the landscape satisfies specific sustainability objectives and meets relevant criteria to assure long-term environmental support of the activities under consideration. The focus on appropriateness also suggests that in the scenario process, consideration is given to spatial organization and critical spatial relationships that will enable the selected scenarios to guide effective decision-making. Scenario planning, therefore, offers a means to connect the technical aspects of planning with the participatory activities that become realized in the scenario. Blending these contrasting aspects of the planning process together produces a systematic way to think and act creatively about the future (Chakraborty and McMillan, 2015). Planning with the scenario also encourages the inclusion of qualitative inputs into forecasting, providing a mechanism to involve nontechnical stakeholders in planning and creates a more inclusive and organized approach for managing situations punctuated by high levels of uncertainty. Although future oriented, scenario planning is not strictly focused on explicit predictions of the future, rather it serves as a method for visioning futures that are not easily explained or estimated by relying solely on trends and assumptions. For scenario planning to be effective, it must be integrated into the wider planning process (Chakraborty and McMillan, 2015). Recognizing that sustainability involves the exploration of uncharted, complex futures, the innovation and theory testing inherent to planning contributes to the need for long-term, multiscalar transformations to redirect current development agendas.

Scenario planning consists of six interacting stages that begin with the identification of a motivating problem (Figure 7.3). The problem, as defined, functions as the focal point for assessment, which directs attention on the examination of alternative solutions. The stages in this process have been summarized by Peterson et al. (2003):

- *Identification of a focal issue*: A focal issue emerges from negotiation among participants in the planning process. They reflect the important aspects of the future that are knowable and can be controlled with a scenario.
- *Assessment*: The focal issue is used to assess the individuals, institutions, ecosystems, and linkages that define the active planning system. Assessment requires a detailed examination of the focal issue in relation to the complexity defining the system and the world views of the actors in the planning process.

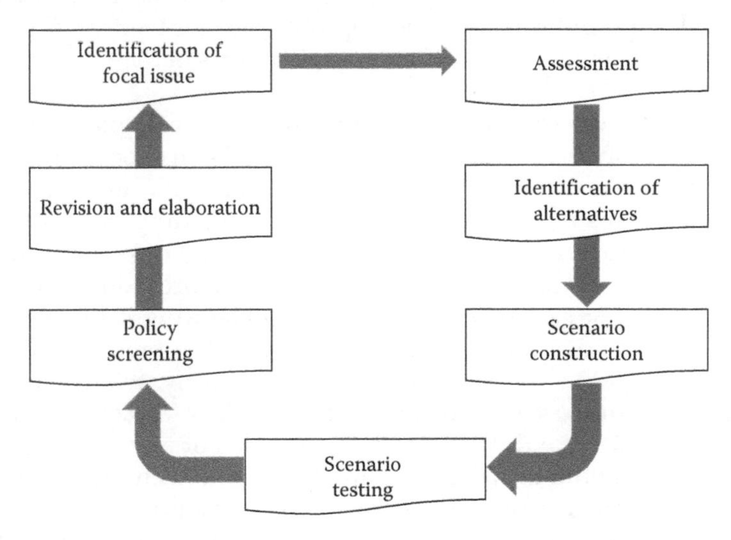

FIGURE 7.3
The scenario planning approach.

- *Identification of alternatives*: A central feature of any planning process are the alternatives that can be defined that are capable of achieving a given goal. Alternatives should be both plausible and relevant, representing a path forward shaped by the interaction of existing dynamics and possible future events. A set of alternatives can be explained by selecting a sample of uncertain or uncontrollable driving forces that push the boundaries of how a future might be conceptualized.
- *Building scenarios*: Following the planning paradigm, scenarios are created based on the understanding gained during assessment. The scenarios serve to elaborate the alternatives and covert the alternatives into dynamic stories by incorporating a credible range of external forces, actors, and responses.
- *Scenario testing*: Testing concentrates attention on the question of scenario consistency. Consistency can be tested in several ways, such as through quantification, expert opinion, or against other scenarios. The quantification approach may involve the application of simulation models of other devices to compare how change may manifest with the system under investigation.
- *Policy screening*: From the set of selected scenarios, each may be used to evaluate, analyze, and formulate specific policy directives. Perhaps one of the more fundamental uses of the scenario in a planning role is to examine how existing policies would fare under possible future conditions. Alternatively, the properties of or the actions defined in the policy can be evaluated in the scenario in relation to their long-term performance. Screening is employed as a means to identify logical traps and gaps in reasoning and consider the feasibility of a policy instrument under different explanations of the future.

In the search for environmental sustainability, scenario planning should enhance the ability of society to manage change. With a focus of the transition to sustainable agendas, decision can be made, policies amended, and management plans implemented to direct the planning system toward a more desirable future. Scenario planning, however, risks succumbing to the same pitfalls as other modeling or planning procedures. Without a more considered or cautious view of scenario planning, adherents to this approach risk the possibility of overly weighing the present and overestimating the capacity to control the future. Similarly, reliance on expert opinion may unnecessarily constrain the breath of a scenario, masking hidden assumptions and the potential consensus of a poor choice. Taken together, the potential drawbacks of employing the scenario as a forecasting tool place emphasis on the importance of developing a vision that can be converted into a strategy for realizing a sustainable future. In the concluding chapter of this book, the aspects of planning and "futuring" that contribute to crafting functional strategies to achieve environmental sustainability objectives are discussed.

7.5 Summary

Futures research depends on the ability of the analyst to tell the story that will suggest a possible future. That story takes shape as a scenario and forms the basis to explore the future, guide thinking, and examine prevailing assumptions. In the context of environmental sustainability, the scenario allows for the creation and examination of alternative strategies and the possible sequence of events that may develop over time. Scenario design and analysis proposes alternative futures through a mix of qualitative and quantitative information to evaluate change and underlying assumptions and communicates a "what-if" image that allows both an assessment of trends and an evaluation of uncertainty. The methods employed to construct, test, and apply a scenario introduce a type of rigor that supports a systematic approach to understanding change and process. The scenario, therefore, supplies planning and visioning activities with the possibilities and uncertainties that are critical to the understanding of environmental sustainability.

References

Abbott, J. (2005). Understanding and managing the unknown the nature of uncertainty in planning. *Journal of Planning Education and Research, 24*(3), 237–251.

Amer, M., Daim, T. U., and Jetter, A. (2013). A review of scenario planning. *Futures, 46*, 23–40.

Bishop, P., Hines, A., and Collins, T. (2007). The current state of scenario development: An overview of techniques. *Foresight, 9*(1), 5–25.

Börjeson, L., Höjer, M., Dreborg, K.-H., Ekvall, T., and Finnveden, G. (2006). Scenario types and techniques: Towards a user's guide. *Futures, 38*(7), 723–739.

Bunn, D. W. and Salo, A. A. (1993). Forecasting with scenarios. *European Journal of Operational Research, 68*(3), 291–303.

Chakraborty, A. and McMillan, A. (2015). Scenario planning for urban planners: Toward a practitioner's guide. *Journal of the American Planning Association, 81*(1), 18–29.

De Brabandere, L. and Iny, A. (2010). Scenarios and creativity: Thinking in new boxes. *Technological Forecasting and Social Change, 77*(9), 1506–1512.

Godet, M. (2000). The art of scenarios and strategic planning: Tools and pitfalls. *Technological Forecasting and Social Change, 65*(1), 3–22.

Godet, M. and Roubelat, F. (1996). Creating the future: The use and misuse of scenarios. *Long Range Planning, 29*(2), 164–171.

Khakee, A. (1991). Scenario construction for urban planning. *Omega, 19*(5), 459–469.

Lemons, J. (1996). *Scientific Uncertainty and Its Implications for Environmental Problem Solving*. Wiley-Blackwell, Malden, MA.

Major, E., Asch, D., and Cordey-Hayes, M. (2001). Foresight as a core competence. *Futures, 33*(2), 91–107.

Mietzner, D. and Reger, G. (2005). Advantages and disadvantages of scenario approaches for strategic foresight. *International Journal of Technology Intelligence and Planning, 1*(2), 220–239.

Nassauer, J. I. and Corry, R. C. (2004). Using normative scenarios in landscape ecology. *Landscape Ecology, 19*(4), 343–356.

Peterson, G. D., Cumming, G. S., and Carpenter, S. R. (2003). Scenario planning: A tool for conservation in an uncertain world. *Conservation Biology, 17*(2), 358–366.

Polasky, S., Carpenter, S. R., Folke, C., and Keeler, B. (2011). Decision-making under great uncertainty: Environmental management in an era of global change. *Trends in Ecology and Evolution, 26*(8), 398–404.

Ratcliffe, J. (2002). Scenario planning: Strategic interviews and conversations. *Foresight, 4*(1), 19–30.

Regan, H. M., Colyvan, M., and Burgman, M. A. (2002). A taxonomy and treatment of uncertainty for ecology and conservation biology. *Ecological Applications, 12*(2), 618–628.

Rounsevell, M. D. A. and Metzger, M. J. (2010). Developing qualitative scenario storylines for environmental change assessment. *Wiley Interdisciplinary Reviews: Climate Change, 1*(4), 606–619.

Schwartz, P. (1992). Composing a plot for your scenario. *Planning Review, 20*(3), 4–46.

Swart, R. J., Raskin, P., and Robinson, J. (2004). The problem of the future: Sustainability science and scenario analysis. *Global Environmental Change, 14*(2), 137–146.

van der Heijden, K. (2011). *Scenarios: The Art of Strategic Conversation*. John Wiley & Sons, Chichester, U.K.

Van Hulst, M. (2012). Storytelling, a model of and a model for planning. *Planning Theory, 11*(3), 299–318.

Varum, C. A. and Melo, C. (2010). Directions in scenario planning literature—A review of the past decades. *Futures, 42*(4), 355–369.

Vermeulen, S. J., Challinor, A. J., Thornton, P. K., Campbell, B. M., Eriyagama, N., Vervoort, J. M., Kinyangi, J. et al. (2013). Addressing uncertainty in adaptation planning for agriculture. *Proceedings of the National Academy of Sciences, 110*(21), 8357–8362.

Wilkinson, A. (2009). Scenarios practices: In search of theory. *Journal of Futures Studies, 13*(3), 107–114.

Zapata, M. A. and Kaza, N. (2015). Radical uncertainty: Scenario planning for futures. *Environment and Planning B: Planning and Design, 42*(4), 754–770.

8

Sustainability Strategies

From the outset, the focus of this book has been on the concept of environmental sustainability and the methods available to support its realization in the future. This elusive ideal was defined as a level of human development that could coexist with the natural system in a manner that would not compromise or degrade environmental process or reduce the viability of the natural system as a habitat for all living things. From this perspective, environmental sustainability was characterized as a hypothesis that certain forms of human development and specific trajectories defining human actions in the landscape could be engaged over extended time horizons with no appreciable impact of environmental functioning. The long-term focus, so deeply embedded in this concept, represents a challenge to planning and management decision-making, and a suite of approaches were examined that lend themselves to the task of projecting environmental sustainability theory forward into the future. In this concluding chapter, a selection of the strategies that have developed from theory will be examined. These are the blueprints used to structure sustainability scenarios and reflect the current state of thinking regarding the larger task of directing the transition from the present to a future development pattern that satisfies the ideals of sustainability.

8.1 The Challenge of Change

The concept of change stands at the center of the sustainability question. Recognizing changing conditions, promoting desirable changes to existing patterns, and forecasting the form and consequences of change in the context of natural and human-driven processes on the environmental system continue to generate debate and encourage scientific investigation (Grimm et al., 2008; Matthew et al., 2010; Mannion, 2014). Change implies difference, and at its most basic level, change may be conceptualized as a deviation from an expected condition. This simple explanation suggests that the status of an object or entity, characterized by either measurement or observation, displays a new characteristic—the changing colors of a deciduous forest as the fall season approaches or the lower value of monthly rainfall over a watershed than the value recorded for previous months on record. Change is therefore both intriguing and wonderfully complex. Change through time is a basic attribute of the planet. The changes that have been taking place, and are taking place, vary in form, scale, duration, and areal extent (Hidore, 1996). Furthermore, changes in the environment are typically nonrandom events, but the result of characteristic physical and biological processes operating on the planet that have given rise to discernable patters that have emerged through time. Change is, therefore, dynamic and serves as evidence of a process. Sustainability, in general, and environmental sustainability, in particular, are also processes. To achieve a sustainable environmental pattern at either the local or regional scale, the directed process of sustainability must consider the governing natural dynamics constantly at work with the environmental systems. Complexity is introduced throughout by human activities that induce perturbations within natural processes and create a web of natural and human-centered forces and influences whose outcomes remain uncertain.

Environmental sustainability is the response to human-direct environmental change and represents the desire to realign human (economic, social, and political) decision-making with greater inclusion of and sensitivity to the reversible and irreversible patterns of environmental change on which society depends (Thomas, 2012). As scenarios are developed to explore the long term, the stage on which these pictures of the future are developed is not constant; all things, human and environmental, are in motion. Forecasting the trajectory of policies and decisions is tampered by the realization that those descriptions of the future form over the shifts of both human and environmental forces. The answer to the fundamental question, "is it environmentally sustainable?" will always depend on how process is conceptualized and change articulated in the worldview model. While the nature of environmental change challenges sustainability future-casting, it need not derail the effort. As society's capacity to alter and modify environmental relationships outpaces our sensitivity to those new conditions introduced, any approach that can examine the possible ramifications of an environmental decision has tremendous importance particularly if it can (after Lein, 1997)

1. Raise questions concerning development strategies
2. Provide insight into the relationship between human actions and environmental degradation
3. Assist in establishing priorities in response to societal growth pressures

Change, whether expressed at a global or regional scale, may assume five innate forms or descriptions that inform how process is conceptualized and dynamic systems interpreted: persistent, rhythmic, cyclical, short-lived, and anthropogenic (Hidore, 1996):

- *Persistent change*: Lasting and enduring are synonyms of a constant, unidirectional increase or decrease in a characteristic over time. Persistence suggests a process that will maintain itself indefinitely until it is no longer measurable.
- *Rhythmic change*: Rhythmic environmental changes define regular oscillations. These are often periodic fluctuations that occur at regular, predictable intervals.
- *Cyclical change*: Defining phenomena that repeat at irregular intervals with varying intensities. These irregular oscillations reference processes that repeat but the time interval between episodes deviates from an average. There are short-term oscillations in the environment in which the time element is 1–10 years, intermediate oscillations that range from ten to hundreds of years, and long-term oscillations where the extremes occur at intervals of more than a thousand years.
- *Short-lived changes*: These represent sudden departures from a normal or steady-state average. Such events may span from seconds to days and are generally sporadic relative to time and space. Often deviations from normal conditions define events that disperse considerable energy and can be difficult to predict. Geophysical processes such as earthquakes, tornadoes, and landslides are examples of short-lived events although their scale, frequency, and spatial characteristics vary.
- *Anthropogenic change*: Sustainability science is centered on this aspect of the change continuum. Human activity has displayed a well-documented capacity to alter, introduce, and induce substantial deviations within the earth–environmental–atmospheric system. The scale, intensity, form, persistence, and spatial and temporal extent of anthropogenic change vary. While every organism changes its environment, the human species has modified habitat to an extent that human interference has increased the pace of change events that in some instances are irreversible. Anthropogenic change impacts natural environmental processes through the long term and collective effects of individual decisions whose scope depends on the social, political, and economic systems that humans have developed over time. Natural variability often obscured by anthropogenic effects and anthropogenic change equally confuses a clear determination of the degree to which an observed environmental change may be attributed solely to human activity (Figure 8.1).

The unique feature of the changes described is that they are directional. The patterns they exhibit reveal trends over time, some becoming more pronounced while others accelerating depending on the interplay between human and natural processes. The possibilities and uncertainties that

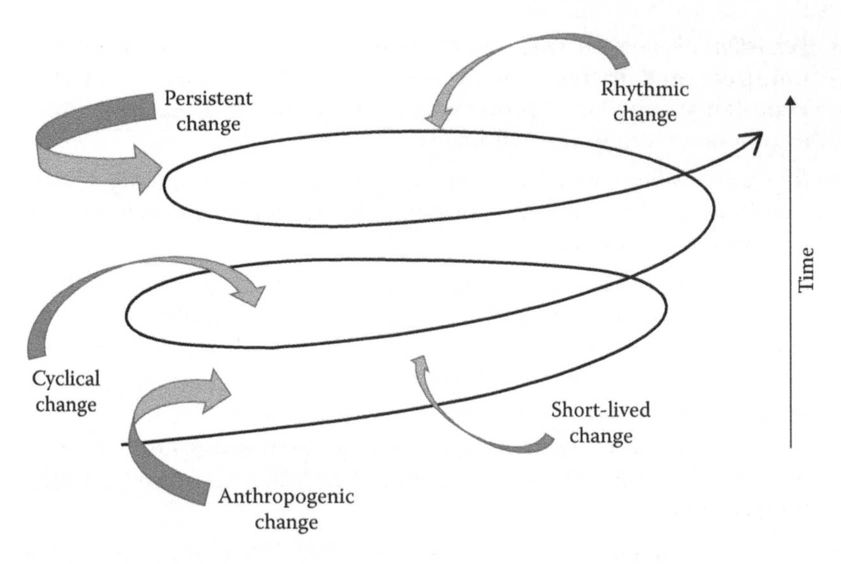

FIGURE 8.1
Characterizing the elements of change.

underscore change creates a dilemma when engaged in sustainability planning and developing specific strategies to achieve targeted outcomes. Given the pervasiveness of change, it is unlikely that the future world will behave as it has in the past. Sustainable management will therefore be required to consider future systems that will not remain the same nor behave as they did in the past. Adopting a more flexible approach to planning with a focus on sustaining the functional properties of the environmental system becomes more complicated by the possibility that the systems involved will be in constant flux. Managing development to respond to and shape change in ways that both sustain and create the same fundamental functions, structures, identities, and feedbacks within the environmental system introduces resilience thinking into the sustainability planning equation.

8.2 Resilience Planning

Resilience may be defined as the capacity of a coupled social-ecological system to absorb an array of disturbances or perturbations and to sustain and develop its fundamental functioning, structure, identity, and feedbacks through the process of recovery and reorganization (Holling, 1973; Folke, 2006). Resilience theory informs sustainability decision-making by focusing inquiry on the recognition and identification of system properties that foster renewal and reorganization following disturbances. The resilience of the environmental system hinges on several conditions, which, if lost or compromised, propel the system to its limits (Chapin et al., 2010). The factors that define resilience include (Folk et al., 2010)

1. *Adaptive capacity*: A condition that explains the capacity of actors, both individuals and groups to respond to, create, and shape variability and modify the state of the system. Adaptability in this context depends on factors such as the degree of biological, economic, and cultural diversity; the capacity of groups or individuals to understand systems behavior; the mechanisms of changes; innovation; and effective governance.
2. *Biophysical and social legacies*: Two factors that define how the existing systems contribute to diversity and offer pathways for rebuilding.
3. *Long-term planning*: A reality that identifies the capacity of actors in the system to consider the long term and confront uncertainty within the context of change.
4. *Response to feedback*: A condition that characterizes the ability to moderate balance between stabilizing feedbacks that buffer the system against stresses or disturbances and the drive to innovate, which generates opportunities to engage change.
5. *Reflective governance*: A property that defines the ability of governance structures to adjust in response to changing needs.

These factors further contribute to three contrasting interpretations of the concept of resilience (Holling, 1986): (1) Engineering resilience that considers the ability of a system to return to an equilibrium or steady state after a disturbance. Here, the resistance to disturbance and the speed by which the system returns to equilibrium become the measures of resilience. The faster the system returns, the more resilient it is. Return time, where efficiency, constancy, and predictability are paramount, is the essential characteristic for a "fail-safe" engineering design. (2) Ecological resilience explains the magnitude of the disturbance that can be absorbed before the system displays a change in structure. This expression of resilience considers the system's ability to persist and adapt while remaining within critical thresholds. (3) Evolutionary resilience, representing a paradigm shift with respect to the ideas of equilibrium and considering the possibility that systems change over time with or without external disturbances. Evolutionary resilience focuses on systems that are inherently chaotic, complex, uncertain, and unpredictable. Therefore, change in a system can result from internal stresses with no proportional or linear relationship between cause and effect. Small-scale changes in a system can, therefore, amplify and cascade into major shifts, suggesting that in some circumstances large interventions may have little effect on the overall behavior of the system.

Integrating resilience thinking into the planning process begins by merging resilience theory with the patterns of human development that drive the sustainability question (Alberti and Marzluff, 2004; Barr et al., 2011; Ahern, 2012). Two integral aspects of planning for resilience can be identified:

1. Strategies to restore or maintain ecological resilience
2. Properties that contribute to resilience in human organizations

Strategies commonly employed to introduce resilience to a planned system include increasing the buffering capacity of the system, managing for processes at multiple scales, and nurturing courses of renewal (Gunderson, 2000). Buffering strategies address resilience in an engineering context and focus on mitigating the effects of unwanted variations in the system. Mitigation solutions attempt to shorten the return time required to arrive back to a desired system state. Where the planning system has been sustained over long periods of time, resilience is increased by managing processes at multiple scales, which could include multiple spatial or temporal domains. Nurturing renewal encourages local communities to co-evolve at time scales matched to the salient characteristics of the processes governing the ecosystem. Overall, planning for resilience concentrates on keystone structuring processes that cross scales, identify sources of renewal, and incorporate multiple sources of capital. However, no single strategy can guarantee resilience. Planning strategies that address a variety of purposes, concentrate on renewal, and encourage deeper understanding of the critical cross-scale interactions are likely to maintain resilience in dynamic environments. Planning with a directed focus of the local and regional contest suggests that different approaches will be required that blend resilience analysis with adaptive planning strategies.

Planning a sustainable future will become more than simply defining a desirable future (Kenny and Meadowcroft, 2002; Dresner, 2008; Barr and Devine-Wright, 2012). It will involve an ongoing decision-analytic procedure that provides a systematic framework for exploring issues and modifying objectives with greater flexibility than is common at present. The core elements of this framework return to the familiar themes of

- Properly formulating the problem
- Specifying feasible alternatives
- Forecasting outcomes attendant to the decision choices
- Selecting criteria for evaluating potential outcomes and consequences

The general sustainability problem typically describes a temporal sequence of decisions over a long, if not infinite, time horizon. The "optimal" action at each decision point will depend on the time and the state of the planning system. The planning goal, therefore, is to develop a policy that prescribes

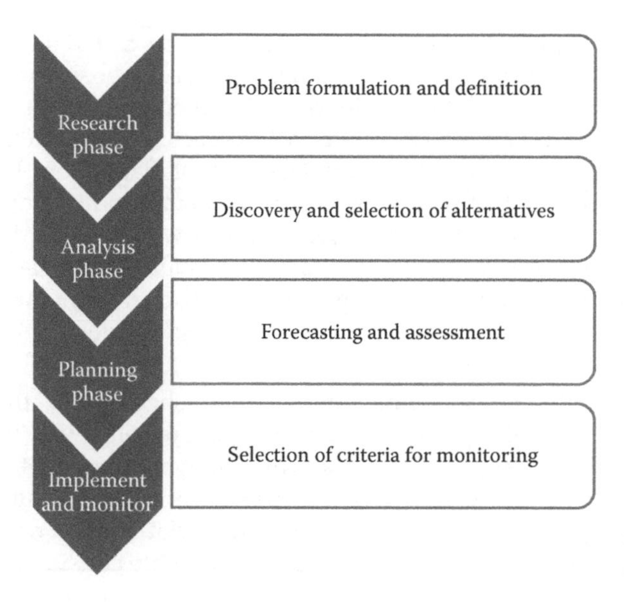

FIGURE 8.2
Features of the sustainability planning method.

a management action for each system state or branch point that is "optical" with respect to the overarching objectives (Figure 8.2). According to this schema, planning assumes a more dynamic role that is called upon to produce a feedback policy to specify "optimal" decisions for possible futures rather than expected future states. In environmental systems characterized by stochastic behaviors, the capacity to remain attuned to the dynamics of the environmental system over an extended time horizon maintains an active response appropriate to the demands imposed by the desire to pursue environmentally sustainable development.

8.3 Policies and Prescriptions

There is growing awareness that a more "scientific" basis for sustainability assessment and planning is needed (Jerneck et al., 2010; Quental et al., 2011; Spangenberg, 2011). It is also well recognized that a one-size-fits-all approach to the future is unrealistic (Grunwald, 2005; Lang et al., 2012). The policies and programs designed to achieve sustainability must take many forms and represent a mix of solutions that are geographically and context relevant. Such strategies also have to adjust over time in response to the inevitable shifts inherent to complex social-ecological systems. A useful guidepost detailing the degree to which substantive policy efforts have been crafted to meet environmental sustainability goals can be found in the pages of existing plans. Traditionally, the regional or comprehensive plan has been the main instrument by which a city or regional entity establishes the goals and objectives that will be called upon to direct future development and address acute social and environmental issues that plague the planning area. In an expanding number of cases, specialized sustainability plans have been adopted that target specific focus areas around which action-based policy objectives are designed. The elements of a sustainability plan may vary; however, several sustainability targets form a common theme throughout that conform to the "three pillars of sustainability" model:

- *Environment*: The planning targets established under this theme focus on (1) natural systems with consideration given to ecosystems and habitats, water and storm water management, air quality, materials waste, and resource conservations; (2) planning and design aspects that detail land use, transportation, and open space/parks and recreation; and (3) energy and climate considerations that concentrate on energy efficiency, alternative and renewable energy resources, green-scaping, and greenhouse gas emissions.
- *Economy*: Targeting economic development centered on green industries, clean technologies, local commerce, and local food systems.
- *Society*: Targeting issues such as societal equity, affordable housing, poverty, human services, community health and wellness, access to health care, public safety, education, civic engagement, and community vitality.

Developing the plan to address these considerations engages a process centered around five planning activities:

1. *Sustainability assessment and review*: This initial phase of the planning process directs the agency or planning jurisdiction to assess its existing sustainability profile. Using a set of metrics to identify key challenges, the assessment review requires research to describe the existing status of environmental, economic, and social equity characteristics. These descriptions serve as a general baseline of existing conditions and a primary source of information to guide the planning process. The assessment may assume one of two basic forms: (a) a quick action assessment that involves a cursory review and inventory of selected sustainability metrics or (b) a comprehensive assessment that will require a more extensive analysis of data regarding factors such as natural resources, land use patterns, population/demographic changes, environmental benchmarks, energy supplies and costs, and environmental hazards. Metrics to quantify these general considerations are typically developed from a sustainability

checklist that highlights a set of criteria that define high-priority issues (Gibson et al., 2013). An example of a sustainability checklist is offered in Table 8.1. Ideally, an assessment, developed from the items provided in the checklist, casts a wide net and permits exploration of a broad set of factors focused not only on the analysis of quantitative data, but also on those qualitative aspects that reflect public concerns and priorities. Once completed, the assessment can be used to define the scope of the sustainability plan and communicate the main challenges that confront the region.

2. *Establishing goals and objectives*: The statement of goals and objective rests at the core of any plan. The sustainability plan presents a "vision" of the futures where the goals and objectives help to direct effects and focus actions to realize that vision. Goals can be an affirmation of an ideal or a response to a problem. In either case, they are subjective and may change with time or circumstance. Goals are used to define the scope of the plan and center around present needs, possible future requirements, community aspirations, and strategic issues. Goals directed at the environmental sustainability question typically form over issues related to greenhouse gas reduction, renewable energy, resource conservation, transportation efficiency, waste management, and environmental stewardship.
3. *Developing the plan*: The sustainability plan outlines the strategies selected to achieve goal attainment. These policies and directives specify a range of measures to implement the plan, including timelines, short-term and long-term milestones, and indicators for measuring progress.
4. *Implementation*: Once the sustainability plan has been adopted, its measures and recommendations are put into action. Implementing the plan requires identifying the various agencies that will be responsible for coordinating and administering its directives. Both tasks involve the allocation of resources as well as an established set of priorities are called upon to order and stage the actions expressed in the plan. A critical issue when implementing the sustainability plan is the need to maintain commitment over the long term. Because environmental sustainability goals are likely to be realized only after the passage of a significant amount of time, momentum can wane and resources initially allocated to the plan may be reallocated to other short-term needs.
5. *Progress tracking*: Progress on the implementation of the environmental sustainability plan can be followed using several reporting options. Perhaps the most useful takes the form of an annual progress report. The annual report provides a means to describe the actions that have been taken to date, the degree to which specific milestones have been met, which next steps are to be anticipated,

TABLE 8.1

Example of a Sustainability Checklist

Urban design components

1 Transportation and mobility

- ☐ Explore and expand alternative transportation options.
- ☐ Understand and evaluate regional connectivity.
- ☐ Ensure equitable access to alternative transportation options.
- ☐ Develop and implement a multimodal plan that includes rail, bus, pedestrian paths, etc.

2 Land use

- ☐ Increase land use efficiency, promoting density and infill development.
- ☐ Promote live/work environments.
- ☐ Coordinate land use plans with regional growth and transportation plans.

3 Green building

- ☐ Implement green building standards (e.g., LEED) for new construction and retrofitting.
- ☐ Establish goals for green development (i.e., ratio of LEED buildings to traditional buildings).
- ☐ Develop incentives to encourage developers to exceed the baseline standards set.
- ☐ Promote adaptive reuse of existing buildings.

4 Housing accessibility, diversity, and affordability

- ☐ Availability of affordable housing, including demographic studies and distribution
- ☐ Evaluation of current housing stock, local, and regional trends
- ☐ Examining and evaluating access to open space, mass transit, etc.
- ☐ Encouraging the rehabilitation of inner city areas

5 Open space, parks, and trails

- ☐ Protect environmentally sensitive areas.
- ☐ Expand parks and recreational opportunities, keeping equitable access in mind.
- ☐ Examine potential linkages to regional trail systems.
- ☐ Provide sufficient urban trail systems to meet the recreational and alternative transport needs for bikers, runners, and pedestrians.

Inputs

6 Alternative energy/energy conservation

- ☐ Increase use of renewable energy.
- ☐ Increase efficiency of existing facilities.
- ☐ Expand usage of alternative fuel vehicles (i.e., city fleets, etc.).
- ☐ Decrease miles driven by city employees (i.e., incentives for mass transit use, modified work week, bike share, etc.).
- ☐ Include consideration for both technological- and policy-based options.

7 Water quality and conservation

- ☐ Increase efficiency (i.e., retrofitting, reuse, harvesting, etc.).
- ☐ Understand the relationship between water supplies/service and land use/growth
- ☐ Perform water auditing.
- ☐ Include consideration for both technological- and policy-based options.

(*Continued*)

TABLE 8.1 (*Continued*)

Example of a Sustainability Checklist

Sustainability-oriented solutions
8 Climate change and air quality
☐ Emission reduction
☐ Pollution prevention
☐ Collaborative partnerships
☐ Incentives
☐ Outreach and education
9 Urban heat island
☐ Examine material usage for roadways, sidewalks, etc., to increase heat absorption.
☐ Increase usage of indigenous vegetation.
10 Waste reduction and recycling
☐ Evaluate effectiveness of existing recycling programs.
☐ Encourage waste reduction (i.e., building operations and construction).
11 Infrastructure efficiency
☐ Coordinate infrastructure and land use plans to maximize infrastructure efficiency.

and the bottlenecks encountered and modifications made to the original plan. The report also supports communication and documents that demonstrate the overall performance of specific programs in relation to key sustainability indicators. The systematic monitoring and tracking provides evidence of success and areas in the plan that call for improvement.

8.4 Examples from the Field

Although environmental sustainability is approached from a variety of contrasting directions, a set of "best practices" has slowly emerged as communities and regional entities have engaged the process of sustainability planning (Berke and Conroy, 2000; Berke, 2002). A review of these plans identifies a set of common themes and strategic initiatives that encapsulate environmental sustainability principles (Figure 8.3).

The plans also define a set of common action items that have been employed to direct the future of the respective planning areas and move those regions toward more sustainable development patterns. While not necessarily exhaustive, the outline gleaned from these plans summarizes three focus areas:

1. Guiding principles
   a. Protect, preserve, and restore the natural environment.
   b. Treat environmental quality, economic health, and social equity as mutually dependent.

Sustainable plan elements		
Community	Built environment	Natural environment
■ Appearance and design ■ Arts and culture ■ Economic development ■ Parks and recreation ■ Community health and safety ■ Fiscal sustainability ■ Diversity ■ Housing attainability	■ Community facilities and infrastructure ■ Housing/ neighborhoods ■ Land use and growth ■ Transportation/mobility ■ Energy conservation and green building ■ Green infrastructure	■ Air quality ■ Environmental resources ■ Environmental hazards ■ Open space ■ Atmosphere and climate change ■ Waste stream reduction/reuse ■ Agriculture ■ Food production and security ■ Renewable energy

FIGURE 8.3
Generic elements common to sustainability plans.

c. All decisions implicative to long-term sustainability.
d. Make community awareness, responsibility, participation, and education as central elements of sustainability.
e. Recognize linkages with the local, regional, national, and global community.
f. Prioritize sustainability issues, addressing the most critical first.
g. Select the most cost-effective programs and policies.
h. Commit to decisions that minimize negative environmental and social impacts.
i. Develop cross-sector partnerships to achieve sustainability goals.
j. Apply the precautionary principle as a complementary framework to guide decision-making.

2. Planning themes
   a. Livable communities
   b. Transportation efficiency
      i. Transportation-oriented land development
      ii. Land use efficient development
      iii. Mass transit mix with alternative modes
   c. Energy conservation
      i. Fossil fuel reduction
      ii. Expansion of renewable sources
      iii. Expansion of energy efficiency programs

      iv. Formation of community energy districts
      v. Greenhouse gas reduction
   d. Waste and materials management
      i. Reduce solid waste generation
      ii. Develop local markets for materials recovery
   e. Agriculture and open space
      i. Protect prime farmland
      ii. Protect priority conservation areas
      iii. Develop local food markets
   f. Water conservation
      i. Reduce consumption
      ii. Improve wastewater treatment
      iii. Reduce impervious surface
      iv. Enhance watershed management
   g. Environmental toxins
      i. Improve air, water, and soil quality
      ii. Reduce exposure to toxins
      iii. Remediate contaminated land
   h. Community health and wellness
      i. Increase access to local health system
      ii. Minimize vulnerability to hazards and disasters
   i. Land character and ecology
      i. Greenspace and greenway development
      ii. Utilization of native plant
      iii. Habitat protection
3. Strategic initiatives
   a. Promote green community and region.
   b. Adopt incentives and regulations to encourage environmental stewardship.
   c. Preserve critical habitat and resource areas.
   d. Recruit and promote green industries and employment.
   e. Develop resource conservation policies.
   f. Promote low-impact site designs.
   g. Promote energy reduction policies and local energy audits.
   h. Revise building and zoning codes to incorporate sustainability goals.

i. Support creation and maintenance of tree canopy to manage local micro-climate.
j. Design future infrastructure to accommodate transportation choice.
k. Promote programs and incentives to reduce solid waste and expand resource recovery.

8.5 Monitoring Sustainability

As plans are implemented, particularly those with extended time horizons, there is a compelling need to monitor the implementation of planning goals against specific indicators of progress. Given the long-term nature of environmental sustainability, an active monitoring system is essential in order to (1) evaluate the extent to which the planning framework has delivered expected outcomes, (2) review and amend specific strategies in relation to new information or changing circumstances, or (3) respond to unanticipated consequences. Therefore, monitoring general trends and conditions that are of environmental and spatial relevance not only informs policy-makers and the public of the realities encountered by the plan, but also provides an active means of responding to and managing the developing patterns and changing agendas.

Monitoring may be defined as the repetitive observation of phenomena over a specific time horizon and across a well-defined spatial extent (Spellerberg, 2005). The general process of monitoring can be understood as the collection, analysis, and interpretation of data concerning the status of the environmental system. When considering environmental sustainability, monitoring attempts to observe both the living and nonliving elements of the landscape and the response of the ecosystem to human interventions. Interventions explain those changes introduced by the activities carried out according to the objectives stated in the sustainability plan. In this regard, monitoring guides decisions on what steps to take in the future and helps gain a more consistent understanding of active environmental processes, providing early warning of negative ecological effects. Although monitoring has been widely advocated in planning and natural resource management (Legg and Nagy, 2006; Nichols and Williams, 2006; Lovett et al., 2007; Lindenmayer and Likens, 2010), its connection to the complexities of environmental sustainability is less well developed (Pediaditi et al., 2010). To be employed effectively in sustainability planning and assessment, monitoring forms the initial component of a tripart design dedicated to the continuous improvement of decision-making and policy actions (Figure 8.4). According to the design, planning entities will, on an ongoing basis, systematically collect and store data for the selected indicators that describe progress toward

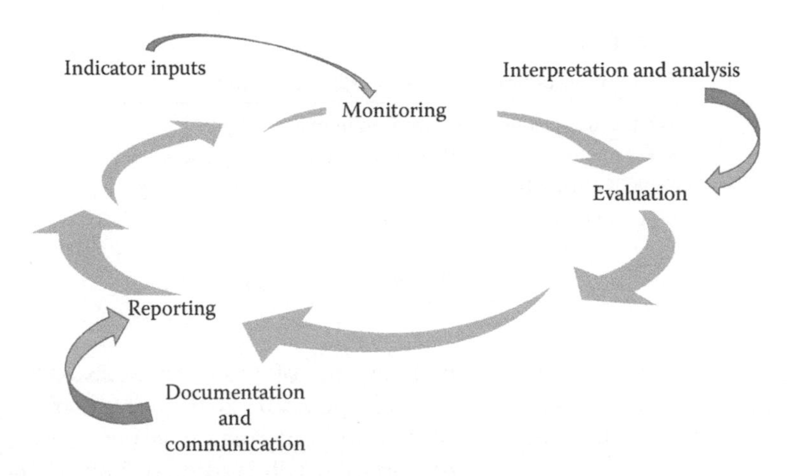

FIGURE 8.4
Features of an environmental sustainability monitoring program.

one or more sustainability goals. The supporting indicators explain the broad economic, social, and environmental measures that feed into plan evaluation. Evaluation concentrates effort on the analysis and interpretation of the indicators against specific benchmarks—limits and triggers (thresholds) established for the region. Reporting introduces the various mechanisms available to formally communicate regional progress. Most typically, progress is outlined within the annual report.

The approach to monitoring and the methods adopted in practice can target three aspects of the sustainability plan: (1) implementation, (2) effectiveness, and (3) evaluation. Implementation monitoring examines the degree to which specific activities have been carried out. Monitoring the implementation process tracks and documents how planning decision have been successful. Monitoring also provides detailed information regarding which management activities have been completed and what further actions may be required for plan implementation to continue. A more challenging form of monitoring, particularly with respect to the ill-defined nature of environmental sustainability, is effectiveness monitoring. Monitoring effectiveness relies on the establishment of clear benchmarks that provide a means to compare against a baseline condition. Positive movement from the baseline according to the suite of selected benchmarks suggests progress. Because effectiveness can be largely subjective, indicators of change, critical thresholds, and specific time frames are essential in order to form a determination of whether desired outcomes have been achieved or whether additional management actions are needed.

Indicators, when placed into a monitoring framework, often assume a binary yes/no notation, implying that a stage in the implementation process has been completed or not. Indicators of this type focus on prompting

TABLE 8.2

Focusing Questions to Guide Plan Review

Focusing Questions to Guide Plan Review
What is happening to the environment and why?
What are the human and environmental consequences?
What is currently being done?
How effective are these measures?
Where are we heading?
What actions could be taken for a more sustainable future?

awareness and communicate which aspect of the plan decision-makers understand and have developed strategies to address. In contrast are indicators that assess which aspects of the sustainability plan have been successful. This class of indicators express both qualitatively and quantitatively conditions of success that can be verified in the data. Feedback from the monitoring program lends support to plan evaluation by encouraging critical assessment of the level to which decisions in the plan have been successful and the degree by which management actions are achieving the desired outcome. Several focusing questions provide guidance for this critical review and direct attention back to the plan and sections in the plan that are underperforming. A selection of these questions is presented in Table 8.2. Answers to these questions can form the basis for plan modification in a maintenance role either to address minor technical changes, make needed corrections to keep the plan up to date, or to amend the plan where substantive changes in objectives, standards, or resource allocations have been identified. The questions presented in Table 8.2 also provide assistance when revising the sustainability plan. Given the time horizon required to achieve a requisite level of environmental sustainability, revisions in a plan are likely as the future redirects initial objectives and presents new challenges.

8.6 Summary

The long-term nature of the sustainability question suggests that as a plan or agenda is implemented, it functions in a dynamic setting that is subject to change and flux. The nature of change within the environmental system and how the system responds to the activities undertaken to achieve a sustainable alignment between human development and environmental process underscores the importance of uncertainty when making plans. To model a better planning solution resilience theory was introduced as a means to inform sustainability decision-making by focusing attention on the properties of the environmental system that foster renewal and reorganization. Introducing resilience thinking into the sustainability planning process merges theory with the patterns of human actions that trigger unsustainable conditions. The planning for resilience approach suggests that sustainable outcomes are achieved by adopting strategies that restore or maintain ecological resilience and encourage forms of human organization that display properties that contribute to more resilience arrangements. Five planning activities can be identified to guide this process, and examples from the current state of sustainability plans suggest a set of common objectives. However, implementing a specific strategy also requires developing an appropriate monitoring system to target the three central components of the sustainability plan: (1) implementation, (2) effectiveness, and (3) evaluation.

References

Ahern, J. (2012). Urban landscape sustainability and resilience: The promise and challenges of integrating ecology with urban planning and design. *Landscape Ecology, 28*(6), 1203–1212.

Alberti, M. and Marzluff, J. M. (2004). Ecological resilience in urban ecosystems: Linking urban patterns to human and ecological functions. *Urban Ecosystems, 7*(3), 241–265.

Barr, S. and Devine-Wright, P. (2012). Resilient communities: Sustainabilities in transition. *Local Environment, 17*(5), 525–532.

Barr, S., Gilg, A., and Shaw, G. (2011). "Helping People Make Better Choices": Exploring the behaviour change agenda for environmental sustainability. *Applied Geography, 31*(2), 712–720.

Berke, P. R. (2002). Does sustainable development offer a new direction for planning? Challenges for the twenty-first century. *Journal of Planning Literature, 17*(1), 21–36.

Berke, P. R. and Conroy, M. M. (2000). Are we planning for sustainable development? An evaluation of 30 comprehensive plans. *Journal of the American Planning Association, 66*(1), 21–33.

Chapin III, F. S., Carpenter, S. R., Kofinas, G. P., Folke, C., Abel, N., Clark, W. C., Per Olsso, P. et al. (2010). Ecosystem stewardship: Sustainability strategies for a rapidly changing planet. *Trends in Ecology and Evolution, 25*(4), 241–249.

Dresner, S. (2008). *The Principles of Sustainability*. Earthscan, New York.

Folke, C. (2006). Resilience: The emergence of a perspective for social–ecological systems analyses. *Global Environmental Change, 16*(3), 253–267.

Folke, C., Carpenter, S. R., Walker, B., Scheffer, M., Chapin, T., and Rockstrom, J. (2010). Resilience thinking: Integrating resilience, adaptability and transformability. Retrieved from http://www.treesearch.fs.fed.us/pubs/42598 (accessed June 6, 2016).

Grimm, N. B., Faeth, S. H., Golubiewski, N. E., Redman, C. L., Wu, J., Bai, X., and Briggs, J. M. (2008). Global change and the ecology of cities. *Science, 319*(5864), 756–760.

Grunwald, A. (2005). Conflicts and conflict-solving as chances to make the concept of sustainable development work. In: Wilderer, P., Schroeder, E., and Kopp, K. (Eds.). *Global Sustainability: The Impact of Local Cultures* (pp. 107–122). Wiley-VCH Verlag GmbH & Co. KGaA, Weinheim, FRG.

Hidore, J. J. (1996). *Global Environmental Change: Its Nature and Impact*. Prentice-Hall Inc, Englewood Cliffs, NJ.

Holling, C. S. (1973). Resilience and stability of ecological systems. *Annual Review of Ecology and Systematics*, *4*, 1–23.

Holling, C. S. (1986). The resilience of terrestrial ecosystems; local surprise and global change. In: Clark, W. C. and Munn, R. E. (Eds.). *Sustainable Development of the Biosphere* (pp. 292–317). Cambridge University Press, Cambridge, U.K.

Gibson, B., Hassan, S., and Tansey, J. (2013). *Sustainability Assessment: Criteria and Processes*. Routledge, New York.

Gunderson, L. H. (2000). Ecological resilience—In theory and application. *Annual Review of Ecology and Systematics*, *31*, 425–439.

Jerneck, A., Olsson, L., Ness, B., Anderberg, S., Baier, M., Clark, E., Hickler, T. et al. (2010). Structuring sustainability science. *Sustainability Science*, *6*(1), 69–82.

Kenny, M. and Meadowcroft, J. (2002). *Planning Sustainability*. Routledge, New York.

Lang, D. J., Wiek, A., Bergmann, M., Stauffacher, M., Martens, P., Moll, P., Swilling, M., and Thomas, C. J. (2012). Transdisciplinary research in sustainability science: Practice, principles, and challenges. *Sustainability Science*, *7*(1), 25–43.

Legg, C. J. and Nagy, L. (2006). Why most conservation monitoring is, but need not be, a waste of time. *Journal of Environmental Management*, *78*(2), 194–199

Lein, J. K. (1997). *Environmental Decision Making: An Information Technology Approach*. Blackwell Science, Malden, MA.

Lindenmayer, D. B. and Likens, G. E. (2010). The science and application of ecological monitoring. *Biological Conservation*, *143*(6), 1317–1328.

Lovett, G. M., Burns, D. A., Driscoll, C. T., Jenkins, J. C., Mitchell, M. J., Rustad, L., Shanley, J. B., Likens, G. E., and Haeuber, R. (2007). Who needs environmental monitoring? *Frontiers in Ecology and the Environment*, *5*(5), 253–260.

Nichols, J. D. and Williams, B. K. (2006). Monitoring for conservation. *Trends in Ecology and Evolution*, *21*(12), 668–673.

Mannion, D. A. (2014). *Global Environmental Change: A Natural and Cultural Environmental History*. Routledge, New York.

Matthew, R. A., Barnett, J., McDonald, B., and O'Brien, K. L. (2010). *Global Environmental Change and Human Security*. MIT Press, Cambridge, MA.

Pediaditi, K., Doick, K. J., and Moffat, A. J. (2010). Monitoring and evaluation practice for brownfield, regeneration to greenspace initiatives: A meta-evaluation of assessment and monitoring tools. *Landscape and Urban Planning*, *97*(1), 22–36.

Quental, N., Lourenço, J. M., and da Silva, F. N. (2011). Sustainability: Characteristics and scientific roots. *Environment, Development and Sustainability*, *13*(2), 257–276.

Spangenberg, J. H. (2011). Sustainability science: A review, an analysis and some empirical lessons. *Environmental Conservation*, *38*(03), 275–287.

Spellerberg, I. F. (2005). *Monitoring Ecological Change*. Cambridge University Press, Cambridge, U.K.

Thomas, C. J. (2012). Transdisciplinary research in sustainability science: Practice, principles, and challenges. *Sustainability Science*, *7*(1), 25–43.

Index

F

T

U

V

W